Vedic Mathematics for Students

Vedic Mathematics for Students

Manu Tripathi

PRABHAT
PRAKASHAN

Published by
PRABHAT PRAKASHAN PVT. LTD.
4/19 Asaf Ali Road,
New Delhi-110 002 (INDIA)
e-mail: prabhatbooks@gmail.com

ISBN 978-93-5521-358-7
VEDIC MATHEMATICS FOR STUDENTS
by Shri Manu Tripathi

Edition
2026

Paperback Price
₹ 300.00 (Rupees Three Hundred only)

Printed at
R-Tech Offset Printers, Delhi

His Holiness Jagadguru Shankaracharya
Sri Bharati Krishna Tirthaji Maharaj

This book is dedicated to the most beautiful and respected lady on this earth MAA Smt. Sharda Tripathi. Whatever good work I do in my entire life is just and only because of your great teachings and your blessings!

Foreword

Everyone in this sacred world acquires his or her own wisdom and knowledge through different sources, different experiences, different teachers and mentors. Additionally, each day brings new lessons in life. As there are different sources of knowledge and many different ways to learn, one can enquire: 'What is the ultimate source of knowledge?'

The human mind is always inclined towards logic and towards logical, rational ways of learning and understanding. If, as human beings, we've been given minds that are inclined towards logic and yet the ultimate wisdom and Truth is divine and supreme, there must be a bridge between this Supreme Knowledge and the logical, human mind.

Vedas can be seen as the bridge between the Supreme Truth and human beings because they provide the clearest, deepest way to logically and rationally understand the divine and sacred Truth of the Creator and the creation.

Our holy *Vedas* are like a detailed architectural design of this entire creation and a detailed, complete code to connect with the Creator. The *Vedas* provide all aspects of knowledge necessary for us to navigate our way through our lives as spiritual beings in human bodies, having human experiences. It is up to each individual to cultivate and harvest any crop from the vast and plentiful field known as the *Vedas*.

I'm so glad to know that Shri Manu Tripathi has authored a book on Vedic mathematics as a great gift of guidance for our beloved youth.

Vedic Mathematics provides a quick solution to many time-consuming processes of calculation. These Vedic solutions will surely amaze and inspire you.

Mathematics is a secular, scientific and universal subject. Calculations are held true around the globe, regardless of cultures, creeds, religions or races. The benefits of mathematical equations are available, accessible and applicable to all people, all over the world.

This present book on Vedic mathematics will be highly useful, especially for those who tend to fear mathematics due to its complexity, and of course, it will also be immeasurably beneficial to those who are already adept mathematicians.

I am sure that this great work of Shri Manu Tripathi will be of great benefit to all in helping to further the understanding of Vedic mathematics.

My blessings and best wishes for the success of this great effort.

—Swami Chidanand Saraswati
President and Spiritual Head,
Parmarth Niketan Ashram
Rishikesh, Uttarakhand

Preface

There is a very rich and deep culture of millions of years present in the Indian sub-continent. Many scholars, saints, sadhus, philosophers, *rishis, maharishis,* astronomers, astrologers, *yogacharyas, Vedacharyas, dharmacharyas* and different masters of this universe have given us extraordinary and outstanding heavenly blessings in the form of *Vedas.*

Vedic mathematics is also a fruit of these blessings and awarded to us as God's gift in solving mathematical problems.

The Indian subcontinent's culture and knowledge, which are a divine and celestial treasure trove to enlighten the entire world. By imbibing it into our lifestyle, we can turn our family, society and nation happier by filling it with positive energy and aura. Therefore, following the Vedic ways would be very enriching.

Sacred Knowledge of Vedic Culture should be an inspirational source not only for the Indian subcontinent but for the entire universe. This would help us in turning this world into a better place to live in harmony and synchronisation.

An important part of this sacred Vedic knowledge is Vedic mathematics, which is a much-needed tool for every human being in the world. This very important tool has helped us all the time even in the past to solve the complex puzzles of this intricate system that we are a part of. It should be an essential part of our present and future knowledge system to continue our progress and solve the innumerable mysteries of this universe.

Vedic system is not just knowledge or religion or culture but it is a way of life. Those who come in touch with this knowledge are actually showered by God's blessings. Once you start learning the Vedic system, you will see that your life will start changing not only by way of education but in other spheres of life as well. It will help you in enlightening the soul, aligning your thoughts and leading a balanced life.

—Swami Satish Kumar
Founder,
Bharatiya Sanskriti Sangh

Acknowledgement

This book is a labour of love and during the journey of writing this book, many people have helped me by providing constant encouragement, valuable suggestions and feedback. Their constant support infused me with the energy to complete this journey. Mentioning everybody's name may not be possible but I would really love to show my gratitude and indebtedness to a few people without their help and support it would not have been possible for me to complete this book and get it published.

I am really graced by the God to get the blessings of great Swami Chidanand Saraswati for this work. His kind words about this book are a great source of encouragement and motivation for me.

Vivek Negi, I really appreciate your selfless and restless contribution during this journey, despite you being busy with your interesting startup, Junior Innovation Lab. Your efforts on the illustrations/graphics have added an interesting flavour to the book.

My family is my strength. My loving wife Shipra Tripathi has been a constant source of unwavering support and encouragement for me throughout my personal and professional life. Thank you so much for bearing all the pains during the time I was busy with the book and was not able to give proper attention to my household responsibilities.

It was my elder son, Yatharth Tripathi, who started digitising my notes on Vedic mathematics without my knowledge. One fine Sunday afternoon, when I was lazing around after lunch, he gave me a pleasant surprise by showing me his work. This incident really motivated me to turn my notes into a book. Without his secret initiative, this book may not have seen the light of the day.

My adorable younger son Atharv Tripathi is always there to help with my work and the best part is he is not only learning from me, he teaches so many new things and whenever I was busy with my other responsibilities, he was the one who motivated me to complete this book. My family members are pillars of constant support, love and affection.

Swami Satish Kumarji founder of Bharatiya Sanskriti Sangh was always there to help me during this journey. Many thanks to you for providing the positive energy which enabled me to complete the work.

Dr. Yamuna Mundade, Technical Officer, World Health Organisation, Geneva, Switzerland was very kind to review my work and provide valuable feedback. It is amazing that even though we were sitting miles apart, we could work together on this project through technology.

Lokesh Sharmaji, who started his career in the satellite TV industry in 1998 and helped many media groups to get established, was always there to shower his help and enthusiasm in the writing of this book.

Dr. Diya Sharma, a great scholar and author of *Sankara Vedanta for Environmental Harmony*, guided me at every step in making of this book.

Kewal Garg, producer of 'Andhadhun' movie (Matchbox Production) was the one who recognised my writing skills and appreciated my work which really motivated me to go ahead with the book.

Inder Singh Negiji, retired. GM of ONGC and a scientist, provided valuable inputs in the form of analytical observations and constant support while writing this book.

Shrey Tyagi, my nephew, offered his support and did the tedious task of proofreading the manuscript. His dedicated efforts and youthful enthusiasm helped me in making the necessary corrections.

I am really grateful to Gaby (Martha Gabriela Ibarra Maya), a teacher at Cancun, Mexico for taking the initiative to translate my manuscript into Spanish. Gaby, your husband Ruben Bernal Ongay and your entire family are always my strength.

Last but not the least, special thanks to Sanjayji for boosting me to reach this moment. As usual, he is always there to give his helping hand whenever needed.

Saurabh Tripathi, Achyut Pathak, Sameer Chougaonkar, Samrat Mishra, Mudit Goel, Lokesh Pant and many other friends and well-wishers were the support pillars in completion of this book. I am really thankful to all of you.

My Spanish friends helped me and guided me in doing my projects and book writing in a smooth way; Erika Granillo is my translator who also translated many of my notes and articles from English to Spanish. Cris Carillo and

Fernando Gonzalez were my friends who made my project with the Education Ministry of Quintana Roo possible. Roberta Chan always helped me to organise my seminars in universities. My dear friend Yosef Shabat, your immense support and helping hand in Mexico can never be forgotten. Jose Baltazar Dzul Zum, a mathematics teacher in Cancun is always a good friend and support. Clau Morgan, principal of one of the great schools of Mexico, helped immensely. Michael Arevalo always supported me as a very good friend.

Finally, I am extremely grateful to all my students and participants at my seminars, workshops and webinars for your valuable feedback and the results shown after learning these techniques prodded me to write this book. Without them, this entire endeavour would have remained inadequate and incomplete.

In case I have missed out on anyone, I hereby express my sincere apologies.

Introduction

Vedic Mathematics, which is the most effective, fun, fast, and magical way of solving Mathematics, was founded by His Holiness Jagadguru Shankaracharya Sri Bharati Krishna Tirthaji Maharaj of Govardhana Matha, Puri (1884-1960). He was a great scholar with six masters degrees in Sanskrit, History, English, Philosophy, Mathematics and Science. His Holiness Jagadguru Shankaracharya Sri Bharati Krishna Tirthaji Maharaj gained this exceptional knowledge from one of the four *Vedas*, the latest one being the *Atharva Veda*. The definition of *Veda,* as given by Sri Shankaracarya himself is quoted below:

"The very word, Veda has this derivational meaning, i.e. the fountainhead and illimitable store-house of all knowledge. This derivation, in effect, means, connotes and implies that the Vedas should contain within themselves all the knowledge needed by mankind relating not only to the so-called 'spiritual' matters but also to those usually described as purely 'secular', 'temporal' or 'worldly' and also to the means required by humanity as such for the achievement of all-round, complete and perfect success in all conceivable

directions and that there can be no adjectival or restrictive epithet calculated to limit that knowledge down in any sphere, any direction or any respect whatsoever."

The *Vedas* that exist today are four in numbers:

RIG VEDA
SAM VEDA
YAJUR VEDA
ATHARVA VEDA

Vedic Mathematics is much faster way of calculation than through regular Mathematics, with 'zero error' and inbuilt checking system. One very interesting fact is that except Vedic Mathematics, all other systems related to Mathematics have limitations. With the help of Vedic Mathematics, we utilise both the hemishperes of our brain (left and right). High speed Vedic Mathematics algorithms are used by NASA, Intel, Microsoft and many other I.T. companies. Vedic Mathematics has 16 simple word formulas, called *sutras*, which are really easy to master and apply.

Global Perception on Vedic Ways!

I really feel like talking about two of my outstanding experiences about how this world looks at Indian culture (Bharatiya Sanskriti).

During one of my memorable workshops in Ecuador, while I was discussing how to memorise things quickly with some references and associations, a lady stood up and asked how one could learn a new language and how had I learnt Spanish so quickly.

I replied, pointing at my shirt, "Well, there are always some associations and references involved with learning skills; for example, how would you say this in Spanish?"

She replied, "Kameeza."

I said, "We call it kameez in Hindi."

Then I pointed towards a table and asked, "What is this called in Spanish?"

She replied, "Meza."

I responded, "In Hindi, we call it mez."

She became excited and asked, "This means your language is driven by our language?"

I replied very calmly, "On the contrary, all the languages of this world are derived from my language, which is Hindi."

Suddenly one boy from Venezuela got up from his chair and said, "I beg to differ from your statement that all the languages are derived from your language."

I wasn't expecting this argument and thought that maybe they were taught something different in their countries. I was curious to know what he had in his mind.

To my surprise, he said, "Your language is Hindi but Sanskrit is the mother of all the languages."

For me, it was a proud moment as since my childhood I had been told that 'Sanskrit is the mother of all the languages'.

On another occasion, during one of my educational projects in Cancun, Mexico, I faced some problem because of which I was not in a very amiable mood. I went for a haircut in a salon intending not to talk to or meet anyone.

The chubby and over-friendly barber welcomed me with a smile and started talking in Spanish as he started giving me a haircut. Even though I was able to understand what he was saying, I was in no mood to converse, so I pretended that I did not know Spanish.

After realising the same, he asked me in his broken English, "Hey man, where are you from?"

I was not in a mood to answer him as I was getting irritated with his non-stop monologue. My mind was not at peace and I wanted to drown myself in silence. However, not replying to such an enthusiastic fellow seemed inappropriate, especially when he was trying to be friendly.

I simply answered, "India."

He became excited, "Tu Indu (you are an Indian)! Wow!"

In many countries, people refer to a person from India as Indian and every Indian is called a Hindu. 'Indu' is not about religion or caste but Indians are known as Hindu in entire Latin America as H is silent in Spanish language and so they pronounce it as 'Indu'.

I was unable to fathom the reason for his excitement on learning about me being an Indian. Even though his excitement surprised me, I was still in no mood to talk.

He kept on talking non-stop, "But man, what is the problem? You look not happy; why?"

I was getting impatient with him as he was pestering me with his questions while I wanted to relax in silence. Even though I knew that this person would not understand what was going on in my head, I went ahead and told him, "Okay, if you want to know, I came here to Mexico to conduct seminars on Vedic Mathematics and memory improvement. After some early successes, now I am facing many rejections. I do not know many people here. My friends, who invited me here, are not willing to invest any more money to market my seminars. As this is a new country for me, I do not have many contacts who would sponsor or book a venue to conduct media-related activities. That's why I am disappointed and want to return to my country."

He listened to me quietly and patiently. After I finished, he said, "Cool, man! It is not a problem. You know, my friend, the entire world is looking at China for business and the entire world is looking at India for acquiring knowledge and education. You are doing a great job; I will help you."

He prepared one Facebook page in Spanish and circulated it among his contacts. He went all out and contacted many of his friends to help me. As a result, I got an excellent response and conducted many seminars. He even refused to take money for the haircut!

Even now there are many friends, students and teachers who keep in touch with me and want me to organise more seminars and workshops in their country. Hats off to my dear friend, YOSEF SHABAT!!

It has become our habit to look outwards for knowledge and acceptance, while the fact of the matter is that the majority of the world looks upon India as a centre for knowledge. India is looked upon as a place for its Vedic ways where knowledge is not broken down into different subjects, but is seen holistically. For example, outside India, science and philosophy may be considered unrelated subjects but in India, both these subjects have always been interlinked.

It gives me immense pleasure and makes me feel proud of our culture when I see such deep respect for our Vedic culture during my interactions with different people.

Playing with Dots
Sixteen *Sutras* of Vedic Mathematics

(With Meanings in English)

1. ***Ekadhikena Purvena***: By one more than the previous one
2. ***Nikhilam Navatscaramam Dasatah***: All from 9, and last from 10
3. ***Urdhva-Tiryagbhyam***: Vertically and crosswise
4. ***Paravartya Yojayet***: Transpose and adjust
5. ***Shunyam Samyasamuccaye***: When the sum is same, that sum is zero
6. ***(Anurupye) Sunyamanyat***: If one is in ratio, the other is zero
7. ***Sankalna-Vyavakalanabhyam***: By addition and by subtraction
8. ***Purnapuranabhyam***: By completion or non-completion
9. ***Chalan Kalanbhyam***: Differences and similarities
10. ***Yavadunam***: Whatever is the extent of its deficiency

11. ***Vyastisamastih***: Part and whole
12. ***Shesanyankena Caramena***: The remainder by the last digit
13. ***Sopantyadvayamantyam***: The ultimate and twice the penultimate
14. ***Ekanyunena Purvena***: By one less than the previous one
15. ***Gunitasamuccayah***: The product of the sum is equal to the sum of the product
16. ***Gunakasamuccayah***: The factors of the sum are equal to the sum of the factors.

□

Vedic Mathematics for Students

When we first begin to learn a new subject, we often see it through the eyes of the teacher, who tells us about it and makes us believe that this subject can only be learned in that way, until one day someone shown us a different, interesting, and easier way, saying that 'this may be a life-changing way for your feelings toward it'.

Learning is a skill and we all believe that the way we learn is the greatest way to learn. So, to see how we approach learning, we'll talk about one of the most fascinating subjects in the world: 'Mathematics'. Before we begin, let's take a look at how we've approached this subject in the past.

We may have previously regarded mathematics as the most difficult subject or our greatest adversary, but I am sure that if you simply read this book without attempting to learn anything, the same subject will become your best friend.

Sanatana Dharma, or Hinduism, is not only one of the oldest religions, but it is also one of the most organised and scientific approach. If you comprehend Hinduism, you would realise that all that is imaginable in this world has already been imagined in Hinduism books. The Vedas are believed to be God's word as passed down to the rishis, and they provide a systematic way to living in this loka (earth) and solving all difficulties and concerns that mankind faces.

So how did Hindus manage to avoid solving mathematics? On the contrary, thousands of years ago, Hindus solved all aspects of mathematics and condensed and summarised the subject under 16 various formulas, sutras, or aphorisms that form the basic foundations of the science known as Vedic Mathematics. It will become a pleasurable experience for you once you begin reading it from the first chapter. The finest part about Vedic Mathematics is that it has ZERO ERROR. It indicates that if you are familiar with Vedic Mathematics, you will never make a mistake in the topic.

The title of this book implies that it is a tool for making friends with this subject, and it is separated into different chapters displaying different approaches, each of which requires some specific methodology to impart a lot of knowledge. If you merely start using these strategies, I am sure you will notice phenomenal changes in your attitude to this fascinating and useful subject.

As I have attempted to compile my limited knowledge of Vedic Mathematics gained from many sources and added some value by offering small chunks of this noble and old wisdom, my humble request to you is that you share this knowledge with others so that the entire community benefits. Such a novel approach to mathematical problem solving should be made available to everyone who seek it.

This book is created in such a way that, rather than

attempting to grasp Vedic Mathematics as a whole, small 'sachets' of approaches are shown and discussed in order to get the reader engaged and feel what a lovely tool it is for solving Mathematical problems.

Sit back and unwind as you embark on your quest to become friends with Mathematics. I am sure that your perspective on mathematics will change, and you will be able to explain how this one book has changed your learning abilities. Welcome to taking advantage of this value-added learning programme that has been provided in an interesting and practical manner.

□

Contents

Placement of Numbers

Mathematics is full of calculations, but did you ever realise that it's a very systematic subject? Just take a look at the given diagram. In the circle given below, start writing numbers beginning from 1 to 9 and then the next numbers above these numbers in a series.

As you can see, the single digit or base numbers are 1 to 9 and the next number 10 is written above 1 and 11 above 2 and so on.

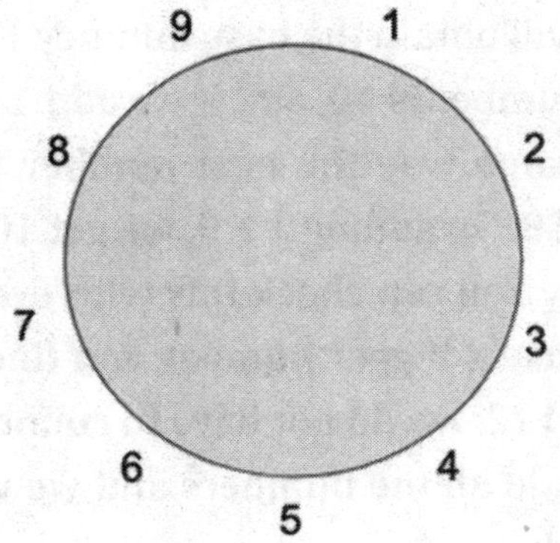

1 10 19 28 (Numbers above 1)

2 11 20 29 (Numbers above 2)

3 12 21 30 (Numbers above 3)
4 13 22 31 (Numbers above 4)
5 14 23 32 (Numbers above 5)
6 15 24 33 (Numbers above 6)
7 16 25 34 (Numbers above 7)
8 17 26 35 (Numbers above 8)
9 18 27 36 (Numbers above 9)

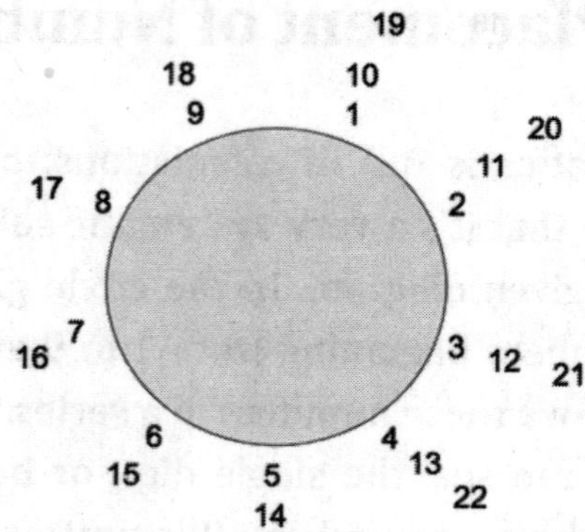

When you total the digits of any number in the outer layers, you will obtain the base number. For instance, above 1, the next number is 10, and if we add 1 and 0, we get 1.

In the same way the next number in the outer layer around 1 is 19. On adding 1 + 9, we get 10; further 1 + 0 = 1

Similarly You can check this with every number, and if we want to use a bigger number and find it's position, say for example 112, we do not have to count it in the circle. All we do is to add all the numbers and we will come to know its base number.

112 = 1 + 1 + 2 = 4

So the base number for 112 is 4.

So the placement of 112 will be above 4.

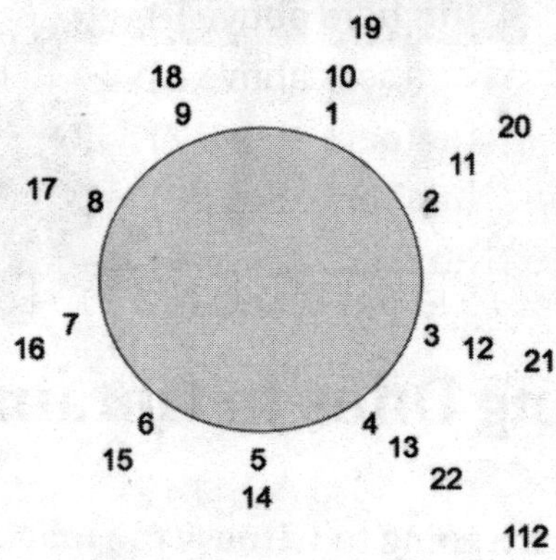

Now if we want to make Mathematics as your good friend, we should start playing with numbers. We can pick any big number and by adding all the digits, we can find the placement of that particular number.

□

Counting Ones to Obtain Squares

Today we are going to witness the magic of mathematics as we start finding squares of 1; like, for example, square of 1, square of 11, square of 111 and so on.

$$1^2 = ?$$

$$11^2 = ??$$

$$111^2 = ???$$

$$1111^2 = ????$$

$$11111^2 = ?????$$

$$111111^2 = ??????$$

$$1111111^2 = ???????$$

Square is one of the basic terms in Mathematics and square of any number means multiplication of the number with the same number. For example, when we say square of 1, it means 1 is multiplied by 1. When we say square of 11, it means 11 multiplied by 11 and so on.

$$1^2 = 1 \times 1$$
$$11^2 = 11 \times 11$$

____

____

To find squares of the above-mentioned numbers in the traditional and conventional way, we would need to employ multi-level multiplication and that is what we have been doing for years. But now is the time to resort to the magical and fun way of Vedic Mathematics and find the correct answer in less time; may be less than 5 seconds. We just have to count 1's in all the above-mentioned numbers to find the squares. This means counting how many 1's are there as digits in that entire number; for example, there is only one 1 in the square of 1, while there are two ones in the square of 11, and 3 ones in the square of 111 and so on.

1^2 = one 1

11^2 = two 1's

111^2 = three 1's

If we know how many 1's are there in the entire number of which we need to find the square, then we just need to start counting from one and up to that number in ascending order and from that peak counting back in the descending order up to 1 and the answer will be that number only. For example, in the case of 111^2, we can see there are three 1's in this entire number. So we count upto 3 in the ascending order and then back down to 1 in descending order. This means it will be 12321 as the answer.

$$111^2 = 12\underline{3}21$$

Now we can find all the squares of numbers having as many 1's, as shown below:

As there is only one 1 in 1^2, so the answer is 1 only.

$$1^2 = 1$$

There are two 1's in 11^2, so the answer is $1\underline{2}1$

$$11^2 = 1\underline{2}1$$

Similarly, there are seven 1's in 1111111^2, so the answer is $123456\underline{7}654321$

$$1111111^2 = 123456\underline{7}654321$$

Now I am sure you too will be able to give the answer to all the above-mentioned one's squares. Here are the answers for your reference:

$$1^2 = 1$$

$$11^2 = 1\underline{2}1$$

$$111^2 = 12\underline{3}21$$

$$1111^2 = 123\underline{4}321$$

$$11111^2 = 1234\underline{5}4321$$

$$111111^2 = 12345\underline{6}54321$$

$$1111111^2 = 123456\underline{7}654321$$

Now write the squares for the numbers given on next page as well, using the same pattern. Trust yourself as you will get all the answers correct and in very less time.

□

Solve and Practice

11111111^2 = ...

111111111^2 = ...

1111111111^2 = ...

11111111111^2 = ...

111111111111^2 = ...

Magic of 25

Let's talk about some more squares. What is the square of 5?

$$5^2 = ?$$

You will get it right, if it is 25.

$$5^2 = 25$$

Now tell me the square of 85.

What will you do to find the answer? You may use the same technique or else use a calculator. I will show you Vedic Mathematics way of finding the squares of any number consisting of 5 as the last digit, e.g. 5, 15, 25, 35, 45, 55, 65 and so on. Just ensure that all the numbers ending with 5, as the last digit, first step is to get the last two digits of the answer, which will always be 25.

$$5^2 = 25$$

This is the first part of the answer. Now what about the complete answer? For that we have to understand the formula of Vedic technique. The formula is: 'by one more than the previous one'. It means that if we apply this formula, we have to know the previous digit, e.g. in 15, 1 is the previous digit and 5 is the last digit. In 25, 2 is the previous digit and 5 is the last digit. Similarly in 35, 3 is the previous digit and 5 is the last digit.

Our formula says 'by one more than the previous one'. This means that when we know our previous digit, we have to add one more number to that. Example, in the case of 25, 2 is the previous digit, and what is one more than 2, it is 3. In the case of 35, what is one more than the previous digit? Yes, it is 4 which in one more than 3. And in the case of 85, it is 9.

We have to multiply the previous digit with one more of that digit.

In the case of 15, we have to multiply 1 with 2.

And in 25, we have to multiply 2 with 3.

In 35, we will multiply 3 with 4.

1. $1\,\underline{5}^2 = ?$

$1\,5^2 = \underline{?}\,25$ ....first step.

In case of 15, we have to multiply 1 with 2

$1 \times 2 = 2$

$\underline{1}\,5^2 = \underline{(1 \times 2)}\,25$. Here 2 is one more than one,

So we multiply 1 with 2

$\underline{1}\,5^2 = \underline{(1 \times 2 = 2)}\,25$ ....second step.

$1\,5^2 = 225$ is the answer.

2. $2\,\underline{5}^2 = ?$

$2\,\underline{5}^2 = \underline{?}\,25$ ....first step.

In case of 25, we have to multiply 2 with 3.

$2 \times 3 = 6$

$\underline{2}\,5^2 = \underline{(2 \times 3)}\,25$

$\underline{2}\ 5^2 = (\underline{2 \times 3 = 6})\ 25$......................second step.

2 5² = 625 is the answer.

3. $3\ \underline{5}^2 = ?$

$3\ \underline{5}^2 = \underline{?}\ 25$................................first step.

In case of 35, you have to multiply 3 with 4

$3 \times 4 = 12$

$\underline{3}\ 5^2 = (\underline{3 \times 4})\ 25$

$\underline{3}\ 5^2 = (3 \times 4 = 12)\ 25$......................second step.

35² = 1225 is the answer.

Once we start acquiring confidence and realise that what we are doing is correct, then solving even tougher numbers becomes much easier. Now we have learnt this particular method of finding squares by using the formula 'by one more than the previous one'. In this manner, rest of the mathematical questions will become easier to solve.

□

Solve and Practice

45^2 = ...

55^2 = ...

65^2 = ...

75^2 = ...

85^2 = ...

95^2 = ...

105^2 = ...

Multiplication

Most people face a problem in calculation or multiplication and use a calculator to find the answer. Now we move to the multiplication section of Vedic Mathematics, where we will see how easily we can do multiplication, even verbally, without using pen-paper or a calculator.

If we have to multiply any two digit number with 11, say, for example, we have to multiply 23 with 11, then:

$$\begin{array}{r} 23 \\ \times\ 11 \\ \hline ??? \\ \hline \end{array}$$

How you were doing it till now? We may first multiply 1 with 3, then multiply with 2. Then in the second row, again we repeat the same process and then do the addition of these two rows to get the answer or we may get confused and re-calculate again.

As per Vedic Mathematics what we do is very simple and easy to learn without getting confused. Here we check the first digit of the number we want to multiply with 11. So the first digit is 2, just write 2 as your first digit of the answer. Now what is the second digit? It is 3, so 3 will come as the last digit of the answer. Finally just do the addition of both the digits 2 and 3.

2 + 3 equals 5. So 5 will come in the middle and will be your answer.

```
          23
        × 11
     ___________
   2 ..................... first step
             3 ........... second step
   2 + 3 = 5 ............. third step
  _________________
   2 5 3 ................. answer
  _________________
```

Let's try again with another example.

```
          34
        × 11
     ___________
   3 ..................... first step
             4 ........... second step
   3 + 4 = 7 ............. third step
  _________________
   374 ................... answer
  _________________
```

In multiplication questions with 11, the process involves only addition. To understand this method more clearly, we take up one more example.

```
   75
 × 11
______

______
```

Here the first digit is 7 and last digit is 5, so we write 7 as the first digit of the answer and 5 as our last digit of the answer. When we do the addition of both the numbers (7 + 5), we get 12 and we write 2 as the middle digit of the answer. We add that carry digit 1 with 7.

```
     75
   × 11
______________
  7 ...............................first step
           5 ............ second step
    7 + 5 = 12 ............ third step
______________
    7 2 5
    1
______________
    825 ........................ answer
______________
```

Adding 1 with 7 gives 8, so our answer for the multiplication of 75 with 11, will be 825.

□

Solve and Practice

57 X 11 = ?

71 X 11 = ?

86 X 11 = ?

98 X 11 = ?

Till now we learned about the mechanism of mathematics works and multiplication of 11 with 2 digit numbers. But multiplication is not always with 2 digit numbers. What if we had to multiply a bigger number with 11? For example, if we have to multiply 12345 with 11

12345 × 11 = ???

Maybe again we revert to the old method. First, we multiply 1 within the first row and then again multiply it with all the numbers for the second row. Now just see, with the help of Vedic Mathematics, if we do the same calculation, what we will do.

```
  12345
×    11
________

________
```

The most interesting part of this technique is that we do not multiply; rather, we do addition.

0 | 1 2 3 4 5 | 0

First, we write the number in the manner given above. We can put a stroke and a zero in front of the number and a stroke and a zero at the end of the number. Now we do the addition from the right to the left, starting from the last digit. Our last digit is 0, and the next digit on the left side is

5, so we add 0 to 5, 0 + 5 and it is equal to 5. So we write '5' as the first step of our answer.

0 | 1 2 3 4 5 | 0

5

In the same manner the second last digit is 5 and we add 5 with the digit immediately on the left side of 5. This is 4. So 5 is added to 4 and the total is equal to 9. So we write 9 as our next step of the answer.

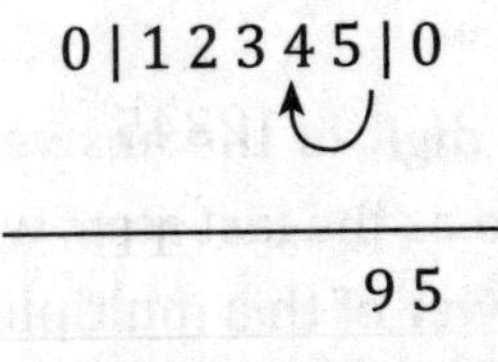

Now as we keep doing the addition, 4 is added to the left side digit 3. Now 4 + 3 is equal to 7.

0 | 1 2 3 4 5 | 0

7 9 5

We continue the same process with the next digit which is 3. We add 3 to the immediate left side digit, which is 2. Now 3 + 2 is equal to 5.

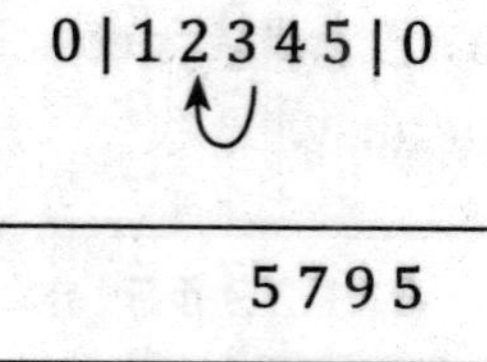

And we go to the next number and add 2 to the left side digit, that is 1. Now 2 + 1 is equal to 3.

0 | 1 2 3 4 5 | 0

3 5 7 9 5

To get the first digit of the answer, we add 1 to 0. So 1 + 0 is equal to 1. So as the last step, we write 1 as the first digit of the final answer of this multiplication sum.

0 | 1 2 3 4 5 | 0

1 3 5 7 9 5

When we want to check how we got the correct answer of multiplication by simple addition, we added last digit to the previous one.

5 + 0 = 5

4 + 5 = 9

3 + 4 = 7

2 + 3 = 5

1 + 2 = 3

0 + 1 = 1

$$\begin{array}{r} 12345 \\ \times \quad 11 \\ \hline \end{array}$$

1 3 5 7 9 5 .... answer

Could this be any easier?

Challenging yourself with new questions is always a great way of finding your own abilities and sharpening your skill. With the help of Vedic Mathematics, we can explore new possibilities in learning a great skill in our life-time.

□

Solve and Practice

31346 × 11

??

22254 × 11

??

25368 × 11

??

46390 × 11

??

Multiplication with 12

In the previous chapter, we talked about multiplication with 11. What about multiplication with other numbers? Now we will see how to multiply any number with numbers from 12 to 19.

Suppose we have to multiply the same number 12345 with 12

12345 × 12

In this case, we have to follow almost the same procedure, with just one change in the rule. Earlier, while multiplying with 11, we were simply adding the previous digit to the next digit, but in the case of multiplication with 12, we have to double the left side digit and then add to the right side digit.

12345 × 12

??

We put a stroke before the first digit: 1, and one stroke after the last digit : 5. We will consider one zero before the starting stroke and one zero after the last stroke, now we repeat the same process the same way we had done for multiplication with 11. The only change in the process is here we multiply the second last digit with 2, before adding it to the last digit.

5 × 2 = 10;   10 + 0 = 10

0 | 1 2 3 4 5 | 0

× 2

0

1

As we have two digits, so we will write 0 and carry 1 for the next step.

We keep on doubling the digit and add to the right side digit.

4 × 2 = 8; 8 + 5 = 13

Now 1 is carried from the last step: 13+1 =14

0 | 1 2 3 4 5 | 0

× 2

4 0

1

Now 1 is carried for the next step.

$$3 \times 2 = 6; 6 + 4 = 10$$

We add the carry over: 10 + 1 = 11

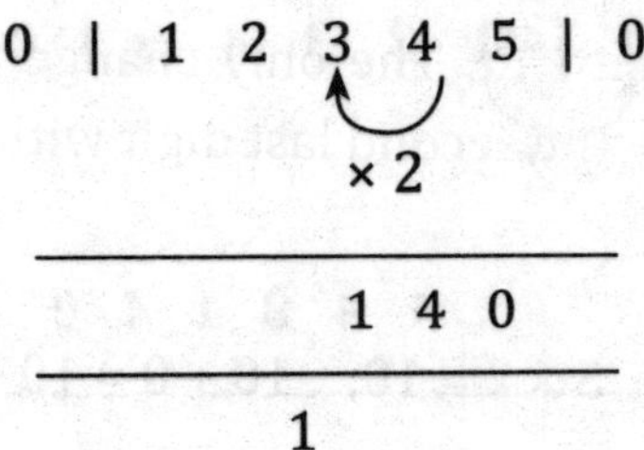

Again there is 1 carried over in this step. We double the left digit and add it to the right side digit to get the next answer.

$$2 \times 2 = 4; 4 + 3 = 7.$$

The carry over in added from the last step: 7 + 1 = 8.

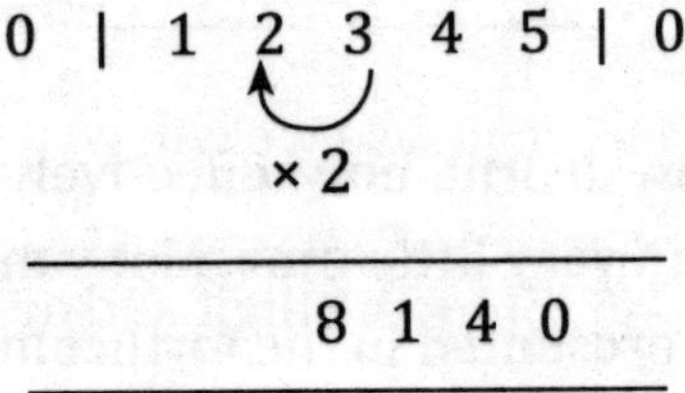

For the next step of the answer, we go further with the same process.

$$1 \times 2 = 2; 2 + 2 = 4$$

0 | 1 2 3 4 5 | 0

× 2

4 8 1 4 0

Now for the final step of the answer:

$$0 \times 2 = 0; 0 + 1 = 1.$$

0 | 1 2 3 4 5 | 0

× 2

1 4 8 1 4 0

So finally, we get the answer and that too with very easy and simple steps.

12345 × 12

1 4 8 1 4 0 ....answer

We have now learnt how effectively we can obtain correct answers in very little time. Many more exciting and fun methods are presented in the forthcoming chapters.

□

Solve and Practice

31346 × 12

??

22254 × 12

??

25368 × 12

??

46390 × 12

??

Multiplication with 13

We observed in the multiplication questions that we can get answers with just simple multiplication and addition. In the same manner, for multiplication with 13, we will triple the left-side digit and add to the right side digit to get the correct answer. For example, if we have take the same numbers, 12345.

$$\begin{array}{r} 12345 \\ \times \quad 13 \\ \hline ?? \\ \hline \end{array}$$

We place a stroke before the first digit: 1, and one stroke just after the last digit, which is 5. We will consider one zero before the starting stroke and one zero after the last stroke, just as we had done for multiplication with 11 and 12. Here, before doing the addition, we multiply the second last digit with 3 and then add it to the last digit.

$$5 \times 3 = 15; 15 + 0 = 15$$

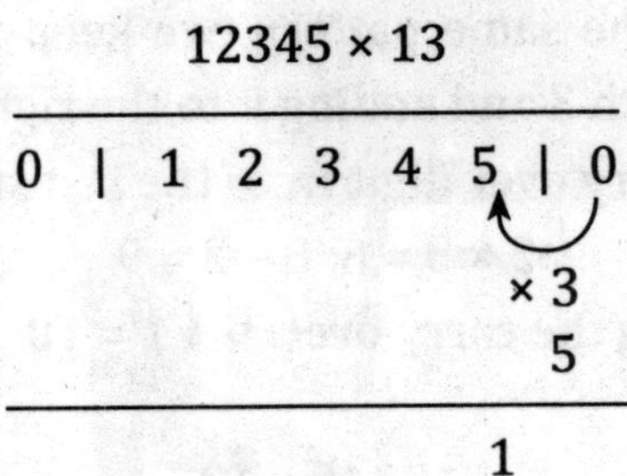

We get 1 as carry over for the next step and repeat the process.

4 × 3 = 12; 12 + 5 = 17

Now add the carry over from the last step:

17 +1 = 18

12345 × 13

0 | 1 2 3 4 5 | 0

× 3

8 5

1

We get 1 as carry over for the next step. So for the next step:

3 × 3 = 9; 9 + 4 = 13

After adding the carry 13 + 1 = 14

12345 × 13

0 | 1 2 3 4 5 | 0

× 3

4 8 5

1

Following the same pattern, we keep multiplying the left side digit with 3 and adding it to the right side digit. We then add the carry over digit from the last step.

$$2 \times 3 = 6; 6 + 3 = 9$$

After adding the carry over: $9 + 1 = 10$

12345 × 13

0 | 1 2 3 4 5 | 0

× 3

0 4 8 5

1

Now in this step also we get 1 as the carry over. So again we multiply the left side digit 1 to 3 and add it with the right side digit, 2. Finally the carry over is 1 which is added to the outcome.

$$1 \times 3 = 3; 3 + 2 = 5$$

On adding the carry over: $5 + 1 = 6$

12345 × 13

0 | 1 2 3 4 5 | 0

× 3

6 0 4 8 5

In this final step to our answer, the same pattern is repeated,

0 × 3 = 0; 0 + 1 = 1.

12345 × 13

0 | 1 2 3 4 5 | 0

× 3

1 6 0 4 8 5

12345 × 13

1 6 0 4 8 5.............answer

□

Solve and Practice

31346 × 13

??

22254 × 13

??

25368 × 13

??

46390 × 13

??

Multiplication with 14

Multiplication is much easier with the help of Vedic Mathematics. We have seen that while multiplying any number with any two digit number, from 11 to 19, we can get the answers very quickly and accurately through simple addition. We can say that the second digit of the two-digit number plays a vital role in this process because before addition, we do multiplication of the left side digit with the second digit of that number. For examples, while multiplying with 12, we first multiply the left side digit by 2 and then add it to the right side digit. In the same way while multiplying with 13, we multiply the left side digit with 3, and then add to the right side digit. So in the same way for multiplication with 14, we multiply the left side digit with 4 and then add that number to the right side digit.

Let's do multiplication with 14 of the same number we have taken so far: 12345 to find how to get the correct answer.

$$12345 \times 14$$

??

We put one stroke before the first digit: 1, and one stroke just after the last digit: 5. We will consider one zero before the starting stroke and one zero after the last stroke, just as we had it for the multiplication with 11, 12 and 13. Before doing the addition, we multiply the second last digit with 4 and then add it to the last digit.

$$5 \times 4 = 20; 20 + 0 = 20$$

12345 × 14

0 | 1 2 3 4 5 | 0

× 4

0

2

We get 2 as carry over for the next step. Now we repeat the process.

$$4 \times 4 = 16; 16 + 5 = 21$$

Now the carry over from the last step is added

$$21 + 2 = 23$$

12345 × 14

0 | 1 2 3 4 5 | 0

× 4

3 0

2

We get 2 as the carry over for the next step.

3 × 4 = 12; 12 + 4 = 16.

After adding the carry over: 16 + 2 = 18

12345 × 14

0 | 1 2 3 4 5 | 0

× 4

8 3 0

1

Following the same pattern, we keep multiplying the left-side digit with 4 and adding it to the right-side digit. Now we adjust the carry over digit from the last step.

2 × 4 = 8; 8 + 3 = 11

After adding the carry over: 11 + 1 = 12

12345 × 14

0 | 1 2 3 4 5 | 0

× 4

2 8 3 0

1

Now in this step also, we get 1 as the carry over. Now again we multiply left-side digit 1 with 4 and add it to the right-side digit 2 and finally add the carry over 1 to the outcome:

1 × 4 = 4; 4 + 2 = 6

Add the carry over: 6 + 1 = 7

12345 × 14

0 | 1 2 3 4 5 | 0

× 4

7 2 8 3 0

In this final step of our answer, the same pattern is repeated:

0 × 4 = 0; 0 + 1 = 1

12345 × 14

0 | 1 2 3 4 5 | 0

× 4

1 7 2 8 3 0

12345 × 14

1 7 2 8 3 0.............answer

The best part of Vedic Mathematics is that once we understand the method, we can obtain the correct answer with different numbers every time we try.

□

Solve and Practice

31346 × 14

??

22254 × 14

??

25368 × 14

??

46390 × 14

??

Multiplication with 19

Now that we have learnt the pattern of multiplication of two-digit numbers with any number, and get the correct answer according to the value of the second digit, we can non try multiplication with 15. We multiply left-side digit with 5 and add it to the right-side digit. For 16, the left-side digit will be multiplied with 6 just as for multiplication with 17 and 18, left-side digit is multiplied by 7 and 8 respectively. The same pattern is repeated with 19 as well.

12345 × 19

??

We put one stroke before the first digit: 1 and another stroke just after the last digit : 5. We will consider one zero before the starting stroke and one zero after the last stroke, the same way we did it for multiplication with 11 or 12 or 13 or 14. Here, before doing the addition, we multiply the second last digit with 9 and then add it to the last digit.

5 × 9 = 45; 45 + 0 = 45

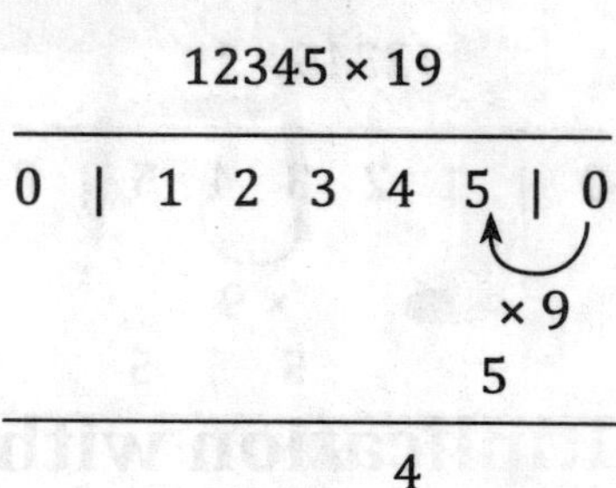

We get 4 as the carry over for the next step. So for the next step, we repeat the process:

4 × 9 = 36; 36 + 5 = 41

Now add the carry over from the last step:

41 +4 = 45

12345 × 19

0 | 1 2 3 4 5 | 0

× 9

5 5

4

As we get 4 as the carry over for the next step and for the subsequent step:

3 × 9 = 27; 27 + 4 = 31

After adding the carry over: 31 + 4 = 35

12345 × 19

```
0 | 1 2 3 4 5 | 0
        ↑‿‿┘
         × 9
        5 5 5
-----------------
      3
```

In the same manner we multiply the left-side digit with 9 and add it to the right-side digit and then adjust the carry over digit from the last step:

$$2 \times 9 = 18; 18 + 3 = 21$$

After adding the carry over, it is: 21 + 3 = 24

12345 × 19

```
0 | 1 2 3 4 5 | 0
      ↑‿‿┘
       × 9
      4 5 5 5
-----------------
    2
```

Now in this step also we get 2 as the carry over. So we multiply the left-side digit 1 with 9 and add it to the right-side digit 2 and finally adjust the carry over 2 by adding it to the outcome.

$$1 \times 9 = 9; 9 + 2 = 11$$

Add the carry over: 11 + 2 = 13

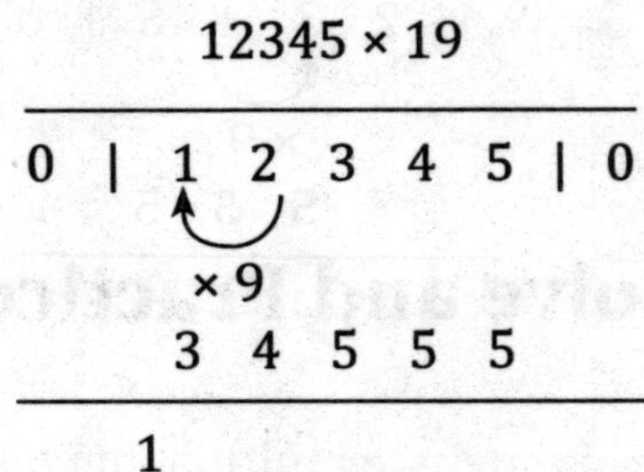

In this final step, the same pattern will be repeated:

$$0 \times 9 = 0; 0 + 1 = 1$$

Add the carry over: 1 + 1 = 2

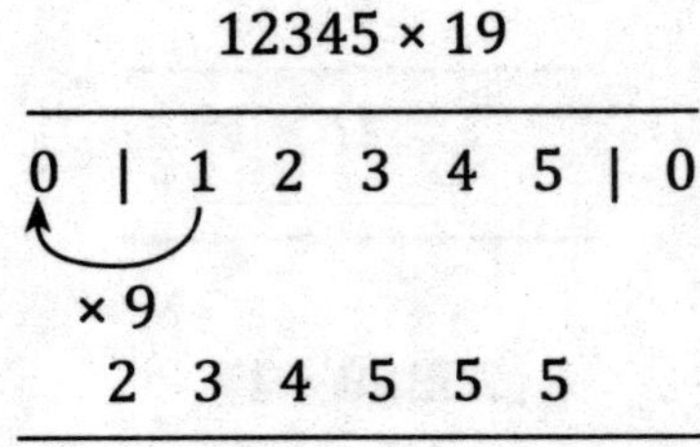

12345 × 19

2 3 4 5 5 5................answer

□

Solve and Practice

31346 × 15

??

22254 × 16

??

25368 × 17

??

46390 × 18

??

Question is the Answer

Let's talk about division now. As we all know, there are four elements of division:

Division

Which is to be divided	:	Dividend
Which divides the dividend	:	Divisor
Number of times divisor, divides	:	Quotient
Whatever is left after division	:	Remainder

So we get ready to learn one more very exciting and magical technique of Vedic Mathematics. We are talking about division of any number by 9. Here we will witness that the question becomes the answer. Let's take a small number to begin with: divide 12 by 9.

$$12 \div 9 = ?$$

Check one very special and interesting feature of division by 9. Here we don't have to do any calculation to find the answer because within the question lies the answer; How? For that we check the dividend. Here the dividend is

12. Now the answer is within the dividend itself as the first digit of the number will become the quotient and addition of both the digits will give the remainder or the answer.

$\underline{1}2 \div 9$

1 → Quotient.........................first step

$\underline{1}2 \div 9$

1+2 = 3 → Remainder............second step

&

1 → Quotient, 3 → Remainder IS THE ANSWER

Let's do it with some other number. Let us divide 23 by 9:

$23 \div 9 = ?$

As first digit of the dividend is 2, so this will be our quotient

2 → Quotient........................ step 1

$23 \div 9$

2+3 = 5 → Remainder............ step 2

and

2 → Quotient, 5 → Remainder is the Answer

Now we take another example of division. Let's divide 46 by 9 with the same technique:

$46 \div 9 = ?$

4 is the first digit in the dividend, so we write 4 as a quotient.

4 → Quotient

And when we do the addition of the digits 4 and 6, we get 10

4 + 6 = 10

10 is a bigger number than the divisor 9, so how can 10 be the remainder?

10 > 9

Now in this case, where the remainder is bigger than the divisor, we extract the sum amount of the divisor. Since in this case, 9 is our divisor, we extract 9 from the remainder 10.

10 – 9 = 1

As we have taken 9 from the remainder, it will be added to the quotient. So the final quotient is:

4 + 1 = 5 → Quotient

After the extraction of 9 from the remainder, which means subtracting 9 from 10, we get 1 as the remainder:

10 – 9 = 1 → Remainder

46 ÷ 9 = ?

5 → Quotient

1 → Remainder is the Answer

Now even with bigger numbers, we don't have to do any calculation to find the answer. We follow the pattern that we used earlier and with just one extra step of addition, we can get the answer. For example, let's say that the dividend is 123. The answer is within the dividend as the first digit of the number will become the first digit of the quotient. On adding this quotient digit to the next digit of the dividend will become the next digit of the quotient. Finally on addition of this quotient digit with the last digit of the dividend will be give the remainder which becomes the answer.

$$123 \div 9 = ?$$

As the first digit of the number is the first digit of the quotient and that is 1, so 1 is our first digit of the quotient.

1

The addition of the first quotient digit to the next digit of the dividend becomes the next digit of the quotient. The first quotient digit is 1 and next digit of the dividend is 2. This means that the addition of 1 and 2 is our next digit of the quotient.

$$1 + 2 = 3$$

13 → Quotient

The addition of the second quotient digit with the last digit of the dividend will become the remainder. The second quotient digit is 3 and the last digit of the dividend is 3. So the addition of both these digits will become the remainder.

3 + 3 = 6 → Remainder

So the final answer is:

123 ÷ 9 = ?

13 → Quotient

6 → Remainder

Another example:

232 ÷ 9 = ?

As the first digit of the number is the first digit of the quotient and that is 2, so 2 is our first digit of the quotient.

2

Addition of the first quotient digit and the next digit of the dividend is the next digit of the quotient. Here the first quotient digit is 2 and next digit of the dividend is 3. This means that the addition of 2 and 3 is our next digit of the quotient.

2 + 3 = 5

25 → Quotient

The addition of the second quotient digit with the last digit of the dividend will become the remainder. The second quotient digit is 5 and the last digit of the dividend is 2 as. So addition of both these digits will give the remainder.

5 + 2 = 7 → Remainder

So the final answer step by step is:

232 ÷ 9 = ?

25 → Quotient

7 → Remainder

Now as we know the pattern and the method to get the correct answer, we can solve more questions on division by 9. Let's take another example with a different number.

762 ÷ 9 = ?

Here the first digit of the dividend is 7. So 7 is the first digit of the quotient

7

And when we add 7 to the next digit of the dividend 6

to get the second digit of the quotient, we get 13

$$7 + 6 = 13$$

But this quotient digit cannot be higher than the divisor (that is 9). This shows that there is one more 9 in this quotient digit. So we extract that and add that to the first quotient digit:

$$(13 - 9 = 4)$$

When we extract one 9 from 13, then 4 remains. It means we can add this that 9 to the first quotient digit (7). Then it becomes 8 and 4 become as the next digit of the quotient.

1
7 4
8 4 → Quotient

Finally, addition of the second quotient digit to last dividend digit gives the remainder.

$$4 + 2 = 6 \rightarrow \text{Remainder}$$

So step by step solution of this question is :

762 ÷ 9 = ?
8 4 → Quotient
6 → Remainder

Now we can understand why the title of this chapter is 'Question is the Answer' because the answer is obtained from the question itself.

□

Solve and Practice

22 ÷ 9 = ___________

32 ÷ 9 = ___________

231 ÷ 9 = ___________

414 ÷ 9 = ___________

2032 ÷ 9 = ___________

Playing with 25

Squares of numbers starting with 5

We have learnt how to find squares of numbers ending with 5. Now we will learn how to get squares of numbers starting with 5. Check how to get squares of numbers starting with 5 i.e. 50, 51, 52, 53 and so on.

Here again is a very easy way of getting the answer but different to what we did for numbers ending with 5.

To find the squares of numbers ending with 5, we put 25 as the last two digits of the answer. But here the numbers start with 5, so the method will be different.

While seeking to get the correct answers of these squares, we will find that even this method is as easy as the previous techniques adopted.

We know that the square of 5 is 25. But here, the numbers start with 5, so should we write 25 as the first answer of all the squares starting numbers with 5? No, but the method is similar.

What is the square of 50 ? We all know it is 2500.

$$50^2 = 2500$$

What we will do here is that as it is a two digit number,

we take the last digit's square at the end, as in 50, the last digit is 0. What is the square of 0? It is 0. So the last two digits will be 00, and the first digit of the question is 5 whose square is 25. But we will not simply write 25 as our first part. What we will do is we add the second digit to 25. Second digit in 50 is 0, so 25+0=25. That is how we get our answer.

$$5\underline{0}^2$$

This means we find the first step of our answer for the square of the number starting with 5 by simply squaring off the second digit of the number and placing that as the last two digits of the answer.

$$\underline{0}^2 = 00$$

Then add the second digit to the square of 5, which is 25.

$$25 + 0 = 25$$

2500..................answer

Let's take another example. What is the square of 51?

$$51^2 = ?$$

Here the second digit is 1, and the square of 1 is 1, so 01 will be our last two digits.

$$5\underline{1}^2$$

$$\underline{1}^2 = 01$$

And then we see that 5 is our first digit so the square of 5 is 25. Then the last digit (that is 1) is added to 25.

$$25 + 1 = 26$$

So 2601 is the answer.

Another example:

$$52^2 = ?$$

Here the second digit is 2. The square of 2 is 4, so 0 and 4 will be our last two digits.

$$5\underline{2}^2$$

$$\underline{2}^2 = 04$$

We find that 5 is our first digit, so the square of 5 is 25. Addition of the last digit (that is 2) to 25, works out to 27.

$$25 + 2 = 27$$

2704 ....answer

Using the same pattern, we can find the correct answers of squares for numbers starting with 5.

$$53^2 = ?$$

Here the second digit is 3, and the square of 3 is 9. So 09 will be the last two digits of the answer.

$$5\underline{3}^2$$

$$\underline{3}^2 = 09$$

We find that 5 is the first digit and the square of 5 is 25. On addition of the last digit (that is 3) to 25, comes to 28.

$$25 + 3 = 28$$

2809 is the answer

The best way to learn and master new knowledge is to challenge ourself to doing more and more practice with new questions and trying to solve those by ourself.

□

Solve and Practice

$54^2 = ?$

$55^2 = ?$

$56^2 = ?$

$57^2 = ?$

$58^2 = ?$

$59^2 = ?$

Difference with Subtract or Add

We have come to a very exciting and very useful part of the techniques where we will learn how to do multiplication of bigger or larger numbers close to the base.

For example we can solve 999 multiplied by 998 in less than 5 seconds and that too with zero error.

$$\begin{array}{r} 999 \\ \times\ 998 \\ \hline ??? \\ \hline \end{array}$$

Before coming to the actual multiplication, let me tell you what is the base. Base, according to Vedic Mathematics, refers to numbers with 1 followed by as many 0s, which we take as base of these numbers:

Base in Vedic Mathematics:

10
100
1000

10000
100000
1000000

____

____

In two steps we can find all the correct answers of multiplication of numbers close to the base. Following are the two steps :

- Multiply vertically .... Step 1
- Addition/subtraction across .... Step 2

Let's take some examples to understand this better. We take some easy and small numbers first to learn the base method.

9 × 8 = ?

When we check both the numbers 9 and 8, we see that both the numbers are close to the base of 10.

10

9
× 8
____
??
____

When we find that both the numbers are below the base i.e. less than 10, we put a stroke just after the number and write the difference between the number and the base. As 9 is 1 below the base 10, so we write –1 after the stroke with 9, and 8 is 2 below the base, so we write –2 after the stroke of 8.

```
10
      9 | -1
    × 8 | -2
  ____________
        ??
  ____________
```

Now we apply the first step rule here to obtain the first part of the answer. The rule says multiply vertically, so we multiply both the digits after the strokes of both the numbers (–1) × (×2)

```
10
      9 | -1 |
             |
    × 8 | -2 ↓
  ______________
             2 ..................Step 1
  ______________
```

One very important point in the base method is that the first part of the answer depends upon the zeros of the base number. The number of zeros decide how many digits the first part of the answer will have. As there is only one zero in base 10, so the first part of the answer will be of single digit. When we multiply (–) with (–) sign, we get answer in (+). That is why the right side answer is 2.

Step 2 says addition/subtraction across. It means we have to do addition or subtraction of either of these numbers with the difference of the other number. As 9 and 8 are our numbers, so we can take either of these numbers. If we take 9, then we have to do addition or subtraction of the difference value of the other number 8. The difference of 8 with the base number 10 is –2, so we subtract 2 from 9.

$$9 - 2 = 7$$

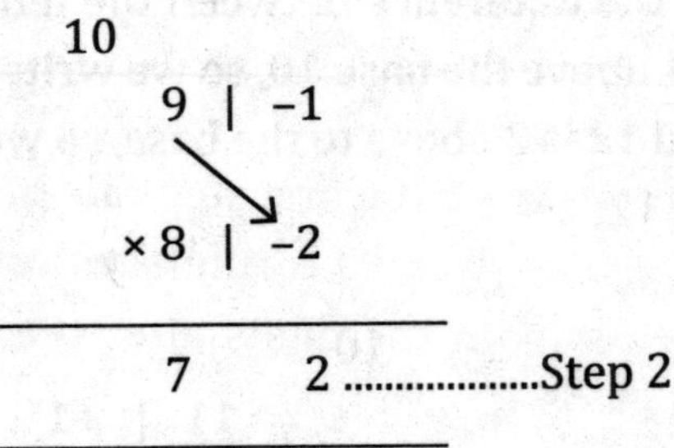

And finally with the help of these two small steps, we get the correct answer in a very short time.

```
   9
 × 8
------
  72 ...............................answer
------
```

Let's take another example: 11 × 12 = ?

Here we find that both the numbers are close to base 10, so we follow the same pattern to get the correct answer.

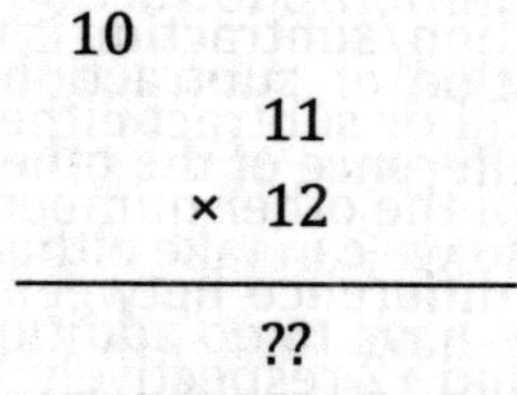

Both the numbers 11 and 12 are above the base number 10, so we put one-one stroke each just after the numbers to mention the difference between the number and the base. As 11 is 1 above the base 10, so we write +1 after the stroke of 11, and 12 is 2 above to the base, so we write +2 after the stroke of 12.

10

11 | +1

× 12 | +2

??

Now we apply the first step rule here to obtain the first part of the answer. The rule says multiply vertically, so we multiply both the digits after the strokes of both the numbers (+1) × (+2)

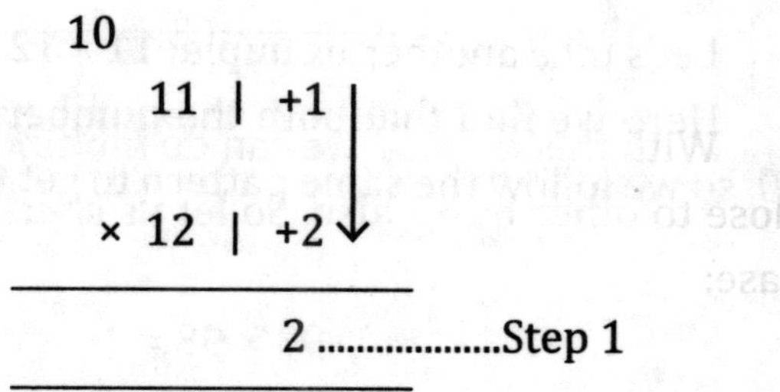

Step 2 says addition/subtraction is to be done across. It means we have to add or subtract either of these numbers with the difference of the other number. Numbers here are 11 and 12 and the difference between both the numbers with the base is +1 and +2 respectively. When we apply this rule with the first number or the second number, we will get the same answer. Let us check:

$$11 + 2 = 13$$

$$12 + 1 = 13$$

This means 13 will be the first two digits of the answer.

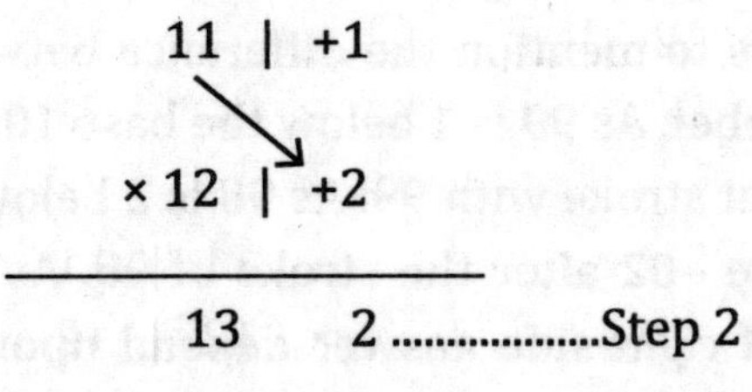

So finally the answer of multiplying 11 × 12 is:

```
  11 | +1
× 12 | +2
-----------
     132 ..................answer
-----------
```

With this method we can do multiplication of numbers close to other bases also. So let us check this with another base:

$$99 \times 98 = ?$$

We find that the numbers 99 & 98 are close to the base of 100, so we take 100 as the base.

```
100
     99
  × 98
__________
     ??
__________
```

Both the numbers, 99 and 98, are below the base number 100. We put one-one stroke each just after the numbers to mention the difference between the base and the number. As 99 is 1 below the base 100, so we write –01 after that stroke with 99. As 98 is 2 below the base 100, so we write –02 after the stroke of 98. As discussed earlier, digits of right-side answer depend upon the zeros of the base number, so when our base is 100 and there are 2 zeros in this base number, the answer of the right side will consist of two digits. Due to this reason, instead of writing –1 & –2 we write –01 and –02 after the stroke.

```
100
     99 | -01
  ×  98 | -02
______________
       ??
______________
```

Now we apply the first step rule here to obtain the first part of the answer. The rule says multiply vertically.

So we multiply both the digits after the strokes of both the numbers (–01) × (–02) and that gives us +02 because a number with –sign is multiplied with another number with –sign, then the answer would be positive.

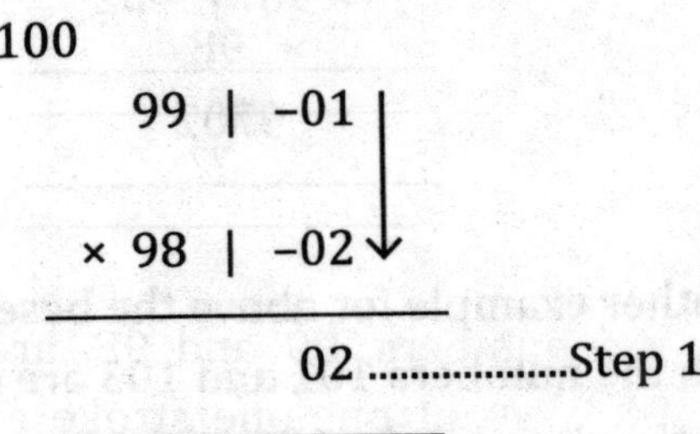

Step 2 says addition/subtraction across. This means that we add or subtract either of these numbers with the difference of the other number. Numbers here are 99 and 98 and the difference of both the numbers with the base is –01 and –02 respectively. We may apply this rule with the first number or the second number, we will get the same answer. Let us check:

99 – 02 = 97

98 – 01 = 97

It means 97 will be the first two digits of the answer.

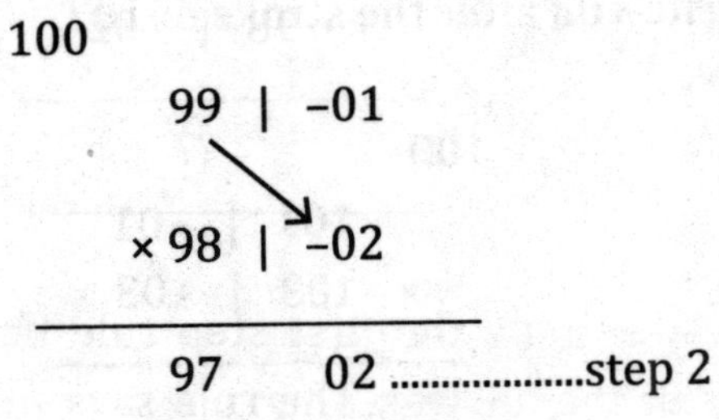

So finally the answer for the multiplication of 99 and 98 is:

```
     99 | -01
   × 98 | -02
  ____________
     9702
  ____________
```

Another example for above the base: 101 × 103 = ?

Both the numbers 101 and 103 are close to the base of 100. That's why, we take 100 as the base.

```
100
      101
    × 103
  ________
      ??
  ________
```

Both the numbers 101 and 103 are above the base number 100. We put one-one stroke each just after the numbers to mention the difference between the number and the base. As 101 is 1 above the base 100, so we write +01 after the stroke of 101. As 103 is 3 above the base 100, so we write +03 after the stroke of 103.

```
100
       101 | +01
     × 103 | +03
  ______________
        ??
  ______________
```

Now we apply the first step rule here to obtain the first part of the answer. The rule says multiply vertically. So we multiply both the digits after the strokes of both the numbers (+01) × (+03) and that gives us +03

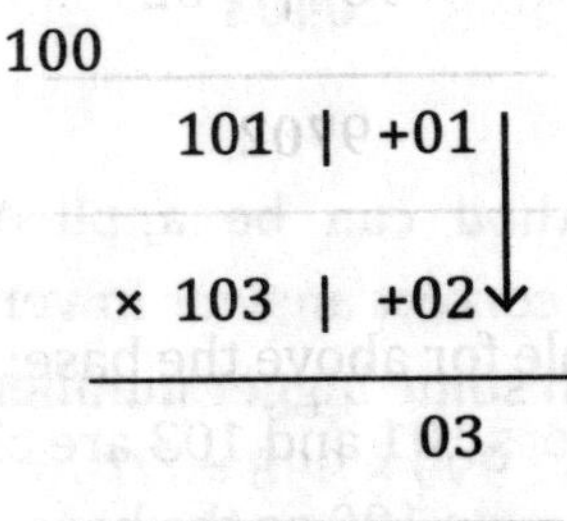

Step 2 says add/subtract a cross which means we have to do the addition or subtraction of either of these numbers with the difference of the other number. Numbers here are 101 and 103 and the difference of both the numbers with the base is +01 and +03 respectively. We may apply this rule with the first number or the second number, we will get the same answer. Let us check:

101 + 03 = 104

103 + 01 = 104

It means 104 will be the first two digits of the answer.

```
  101 | +01
× 103 | +02
-------------
  104   03
-------------
```

So the final answer is:

$$\begin{array}{r|l} 101 & +01 \\ \times\ 103 & +03 \\ \hline \multicolumn{2}{c}{10403} \\ \hline \end{array}$$

This base method can be applied to even bigger numbers to get the correct answer in very little time. Let's try this method with some bigger numbers now:

999 × 998 = ???

$$\begin{array}{r} 999 \\ \times\ 998 \\ \hline ??? \\ \hline \end{array}$$

As both the numbers are close to base 1000, so we will use a base of 1000 only.

$$\begin{array}{l} 1000 \\ \qquad\quad 999 \\ \qquad \times\ 998 \\ \hline \qquad\quad ??? \\ \hline \end{array}$$

Both the numbers, 999 and 998, are below the base by –1 and –2 respectively and as our base here consists of 3 zeros, so we may write –001 and –002 to obtain our right-side answer in 3 digits.

```
1000
        999 | –001
      × 998 | –002
     ______________
          ???
     ______________
```

After applying the first rule of the base method we multiply –001 with –002. So we get +002 as the result and answer on the right side.

```
1000
        999 | –001  |
                    |
      × 998 | –002  ↓
     ______________
              002
     ______________
```

And to get the answer of the first digits we apply the second step, which says add/subtract across with either numbers. We get the same answer:

999 – 002 = 997

998 – 001 = 997

So we can write 997 as the starting digits of the answer.

```
1000
        999 | –001
      × 998 | –002
     ______________
        997   002
     ______________
```

So the final answer for 999 × 998 is:

999 | -001

× 998 | -002

997002 .... answer

After learning these techniques, you can easily understand that Mathematics is not always complicated and even big calculations become easy with Vedic Mathematics.

□

Solve and Practice

13 × 12 = ??

14 × 11 = ??

8 × 8 = ??

14 × 13 = ??

98 × 96 = ??

102 × 104 = ??

98 × 103 = ??

998 × 997 = ??

1002 × 1004 = ??

9999 × 9998 = ??

Square of any Two-Digit Number

We have understood so far how to find squares of numbers starting with 5 and squares of numbers ending with 5. Now, how do we find the squares of numbers without 5? Let us see and learn how to find squares of any two digit number.

The answer of squares of any two digit numbers comes in 3 very simple steps:

- Square of the last digit gives the last part of the answer .... Step 1
- Square of the first digit given the first part of the answer .... Step 2
- Multiplication of both the numbers and doubling it gives the middle part of the answer ....Step 3

Now let's take a look at the first example:

$$62^2 = ?$$

2 is the second digit of this number, which gives the last part of our answer after getting the square of 2.

$6\underline{2}^2$

$2^2 = 4$ .... Step 1

6 is the first digit of this number so we find the first part of our answer after getting the square of 6.

$\underline{6}2^2$

$6^2 = 36$ .... Step 2

Multiply both the numbers and double it for the middle part of the answer. Here numbers are 6 and 2:

$\underline{6}2^2$

$(6 \times 2) \times 2 = 24$ .... Step 3

Now we can check our answer with all the steps combined and adjust the carry over from previous steps to get the final answer:

$\underline{6}2^2$

36 $\underline{2}$4 4

24 is the middle part of our answer, which cannot be of two digits. So we keep 4 as our middle part of the answer and take 2 as a carry over for the first part.

$\underline{6}2^2$

36 4 4

$\underline{2}$

After adjusting the middle number, we get our final answer.

$\underline{6}2^2 = ?$

38 4 4

3844 is the answer.

Let's take another example: $31^2 = ?$

Here 1 is the second digit of this number. We find the last part of our answer after getting the square of 1.

$3\underline{1}^2$

$1^2 = 1$ .... Step 1

3 is the first digit of this number. So we find the first part of our answer after getting the square of 3.

$3\underline{1}^2$

$3^2 = 9$ .... Step 2

Multiply both the numbers and double it for the middle part of the answer. Here numbers are 3 and 1:

$3\underline{1}^2$

$(3 \times 1) \times 2 = 6.$ .... Step 3

Now we can check our answer with all the steps combined:

$3\underline{1}^2 = ?$

9 6 1

961 is the answer.

□

Solve and Practice

$17^2 = ?$

$41^2 = ?$

$34^2 = ?$

$63^2 = ?$

$72^2 = ?$

$84^2 = ?$

Little Up and Little Down

Most of the people in this world keep following one set pattern with one particular subject or one particular topic throughout their entire life, as they believe they cannot improvise or do experiments with that. Actually, they have been taught that particular method only. My motive to write this book is to convey that there can be a possibility to find some different way to tackle a challenging situation. Some of you might have practised the questions given in the work sheets only while some may have tried some more challenging questions on their own. The base method is really ideal and helpful in working out bigger calculations, so much so that we can easily do the multiplications of numbers close to the base, without using a calculator.

Till now we have solved problems with the base method where either both the numbers are above the base or both the numbers are below the base. But what about the case where one number is above the base and another number is below the base? So, now we will learn how to solve problems where one number is above the base and the other number is below the base. For example, we take one number above the base of 10 and one number below to the base 10:

$$11 \times 9 = ?$$

10

$$\begin{array}{r} 11 \\ \times\ 9 \\ \hline ?? \\ \hline \end{array}$$

Here the number 11 is above the base and 9 is below the base number 10. We put one-one stroke each just after the numbers to mention the difference between the number and the base. As 11 is 1 above the base 10, so we write +1 after that stroke after 11, and 9 is 1 below the base, so we write –1 after the stroke after 9.

10

$$\begin{array}{r|r} 11 & +1 \\ \times\ 9 & -1 \\ \hline & \\ \hline \end{array}$$

Now we apply the first step rule here to obtain the first part of the answer. The rule says multiply vertically. So we multiply both the digits after the strokes of both the numbers (+1) × (–1). Here we multiply +ve with –ve, so our answer will be –ve.

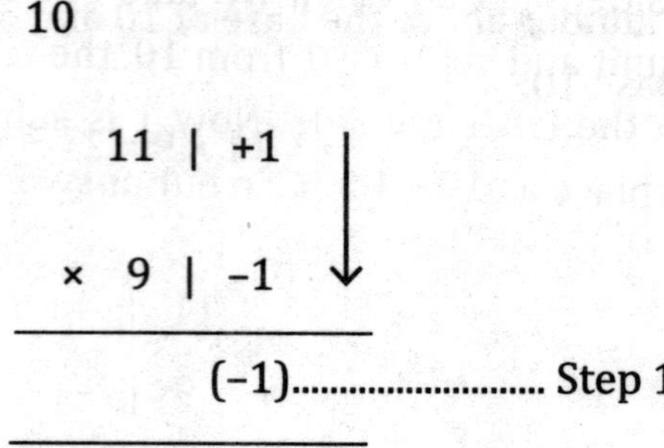

Step 2 says addition/subtraction across. This means we add or subtract either of these numbers with the difference of the other number. Numbers here are 11 and 9 and the difference of both the numbers with the base is +1 and –1 respectively. Now we can apply this rule with the first number or the second number and we will get the same answer. Let us check:

$$11 - 1 = 10$$
$$9 + 1 = 10$$

It means 10 will be placed as the first two digits of the answer.

```
  11 | +1
 × 9 | –1
-----------
 10    (–1)......................Step 2
-----------
```

But this cannot be our answer as the last digit of the answer is with the negative sign. We have to adjust this to get the correct answer. So we take one compliment from the left-side digit. So when we take one compliment from the ten unit and subtract 1 from 10, the remaining digit will be 9 on the left-hand side. Now 1 is subtracted to adjust the unit place as 10 – 1 = 9. So our answer will be 99.

```
      11 | +1
  ×    9 | -1
  ___________
      (→)
    10   (−1)
     9     9
  ___________
```

So finally the answer to multiplication of 11 × 9 will be:

```
      11 | +1
  ×    9 | -1
  ___________
         99 ------------ answer
  ___________
```

To get the correct answer with this method, we have to understand the placement value of numbers. The right-most digit is placed in unit place. This means that the value of this number is in single unit. Next left digit is in ten place, which means the value of this number is in 10s and the next left digit is 100 and so on. That's why when we take a compliment from the left digit, we have to consider the value of that number's place to obtain the correct answer.

Now let's do another sum with one number above and

another below the base to make the method more clear:

101 × 99 = ?

```
100

          101
        × 99
     ___________
          ??
     ___________
```

Here the number 101 is above the base and 99 is below the base number 100. We put one-one stroke each just after the numbers to mention the difference between the base and the number. As 101 is above the base 100, so we write +01 after that stroke (which in 101) and 99 is below the base, so we write -01 after the stroke after 99.

```
100

       101 | +01 |
     ×  99 | –01 ↓
     ______________

     ______________
```

Now we apply the first step rule here to get the first part of the answer. The rule says multiply vertically. So we multiply both the digits after the strokes of both the numbers (+01) × (–01). Here we are multiplying plus sign with minus sign so our answer will be with minus sign.

```
        10

            101 | +01
          ×  99 | –01
        ____________________
                 (–01)........................ Step 1
        ____________________
```

Step 2 says add/subtract across; it means we add or subtract either of these numbers with the difference of the other number. Numbers here are 101 and 99 and the difference of both the numbers with the base is + 01 and –01 respectively. Either we apply this rule with the first number or the second number but we will get the same answer. Let us check:

$$101 - 01 = 100$$
$$99 + 01 = 100$$

This means 100 will be placed as the first two digits of the answer.

```
     100

            101 | +01
                ↘
          ×  99 | –01
     ______________________
            100   (–01).....................Step 2
     ______________________
```

But this cannot be our answer as the last digits of the answer is with negative sign which has to be adjusted to get the correct answer. So we take one compliment from the left-side digit, that is, we take one compliment from the hundred unit and subtract 1 from 100. The remaining digit will be 99 on the left-hand side and one hundred is subtracted to adjust the tens and unit place, that is, 100 – 1 = 99. So our answer will be 99.

100

	101	+01
×	99	–01

→

100 (–01)
99 99

So finally the answer of multiplying of 101 × 99 will be:

	101	+01
×	99	–01

9999 ------------- answer

With the help of this method, we can do multiplication of numbers close to the base, even if they are above or close to the base.

□

Solve and Practice

12 × 8 = ??

14 × 9 = ??

98 × 103 = ??

102 × 97 = ??

98 × 101 = ??

998 × 1002 = ??

1002 × 999 = ??

9999 × 10002 = ??

Just Nines and One Ten

All from 9 and last from 10

Subtraction

This is a very easy and important method of solving subtraction problems that can be used for the base numbers, like 10, 100, 1000, 10000 and so on. This will prove very helpful when we deal with cash exchange.

This chapter is about learning of how to do subtraction from base numbers. Here we do not subtract from the number itself rather we do subtraction from 9's and 10 only, simply subtract all the digits from 9 except the last digit as that has to be subtracted from 10 to get the accurate answer.

For example:

1000 – 625 = ???

Now with the help of our formula, we subtract all the left-side digits from 9 and last digit from 10.

$$\begin{array}{r} 1000 \\ -\ \ 625 \\ \hline \\ \hline \end{array}$$

So we apply this method by subtracting 6 from 9 (9-6= 3), 2 from 9 (9- 2 = 7)) and the last digit 5 from 10 (10-5=5).

$$\begin{array}{r} 9 \\ -6 \\ \hline 3 \\ \hline \end{array} \qquad \begin{array}{r} 9 \\ -2 \\ \hline 7 \\ \hline \end{array} \qquad \begin{array}{r} 10 \\ -5 \\ \hline 5 \\ \hline \end{array}$$

So the answer of 1000 – 625, is 375

1000 – 625 = 375 --------- answer

Another example:

10000 – 4327= ???

Now with the help of our formula, we subtract all the left-side digits from 9 and last digit from 10.

$$\begin{array}{r} 10000 \\ -\ 4327 \\ \hline \\ \hline \end{array}$$

So here 4 is subtracted from 9 (9 – 4 = 5), 3 is subtracted from 9 (9 – 3 = 6), 2 is subtracted from 9 (9 – 2 = 7) and the last digit 7 is subtracted from 10 (10 – 7 = 3).

9	9	9	10
–4	–3	–2	–7
5	6	7	3

So the answer for the number 10000 is 10000 – 4327 which equals 5673.

10000 – 4327 = 5673 --------- answer

□

Solve and Practice

100 – 72 = ??

100 – 38 = ??

1000 – 532 = ??

1000 – 819 = ??

10000 – 7654 = ??

10000 – 8138 = ??

Pole with No Hole

Flag Method–Division

Number to be divided	:	Dividend
Number divides the dividend	:	Divisor
Number of times divisor, divides	:	Quotient
Number left after division	:	Remainder

Now in this chapter we will learn, how to do division by two digit number.

As in the above picture of flag on the pole, we see that the flag-pole comes first and then comes the flag. So in the same manner, in a two-digit number, we consider the first digit as our flagpole and second digit as our flag. To solve division problems with this method, we adopt two steps:

1. Divide by the flag pole.
2. Subtract by (the flag × previous quotient digit).

Example 1:

$$848 \div 31 = ?$$

In 31, we assume 3 as our flagpole and 1 as a flag. So we write 31 as given below:

$$3^1 \left| \; 8 \;\; 4 \;\; 8 \; \right|$$

A very interesting point to consider in this method is that we are using only one digit to divide, i.e. the flagpole so we put one stroke just before the last digit of the dividend. So when we start dividing the last digit, we have to put a decimal in the answer.

$$3^1 \left| \; 8 \;\; 4 \; \right| \; 8$$

Our first rule says that divide by the flagpole. The flagpole of the divisor is 3, so we start dividing the dividend by 3. When we divide 8 by 3, we get 2 as our quotient digit and 2 as the remainder.

3^1	8 4	8	
	2		
	2		Quotient

Now when we check our second step, it says subtract by (the flag × previous question digit). This means that when we check the question, the next number to be divided is 24. So we have to subtract the flag (1) multiplied by the previous quotient digit (2) from 24.

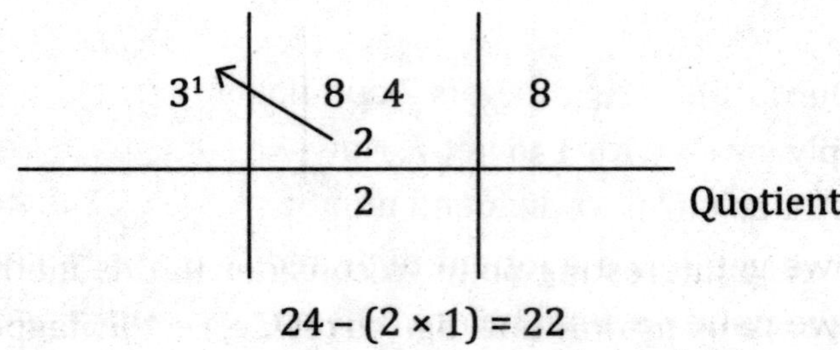

$$24 - (2 \times 1) = 22$$

So when we subtract 2 from 24, we get 22. This means our next dividend number is 22. When we divide 22 by 3, we get 7 as a quotient digit and 1 as the remainder.

3^1	8 4	8	
		1	
	2 7		Quotient

In the same manner we repeat the second step for other dividend numbers. We see that 18 is the next dividend number according to the question but that is not our actual dividend number. Our actual dividend number will come once we repeat the second step.

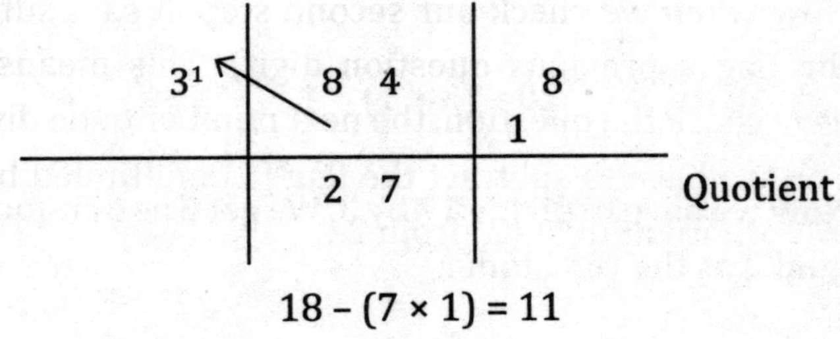

$$18 - (7 \times 1) = 11$$

Our last quotient digit is 7 and flag number is 1, so we multiply seven with 1 to get 7. Now we subtract 7 from 18 to get 11 as our next dividend number. When we divide 11 by 3, we get 3 as a quotient digit and 2 as the remainder. Now we put a decimal before 3 as our quotient digits will come after the decimal only.

3^1	8 4	8 $\;_2$0	
	2 7	3	Quotient

We again repeat the second step and when we multiply the last quotient digit with the flag number, that is, we multiply 3 with 1, we get 3. When we subtract 3 from 20, we get 17 as the next dividend number.

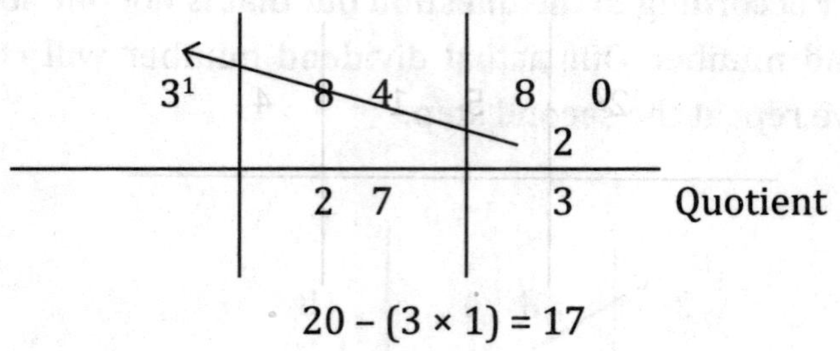

$$20 - (3 \times 1) = 17$$

Now we have to divide 17 by 3. We get 5 as our quotient digit and 2 as the remainder.

3^1	8 4	8 0 0	
		2	Remainder
	2 7	3 5	Quotient

27.35 is the answer

Example 2:

$$514 \div 22 = ?$$

For 22, we assume 2 as our flagpole and 2 as a flag. So we write 22 like a flag:

2^2	5 1 4	

Now we put one stroke before the last digit of the

dividend (see below) to put one decimal at that point in the quotient digit.

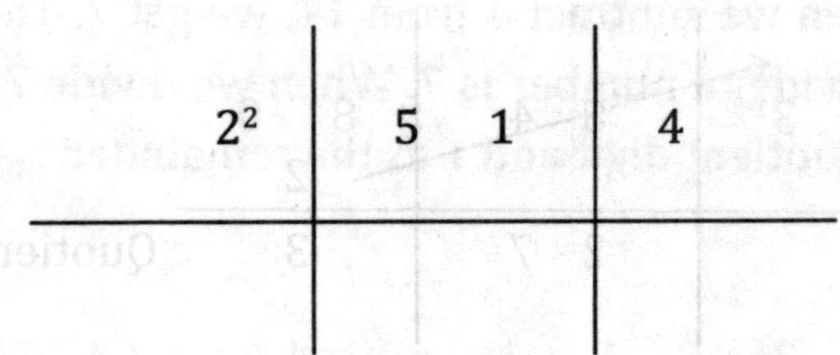

Our first rule says that divide by the flagpole. The flagpole of divisor is 2. So we divide the dividend by 2. When we divide 5 by 2, we get 2 as our quotient digit and 1 as the remainder.

2^2	5 1	4	
	1		
	2		Quotient

Now the second step, says subtract by the flag × previous quotient digit. It means that when we check the question, the next number to be divided is 11. So we subtract the flag multiplied by the previous quotient digit from 11.

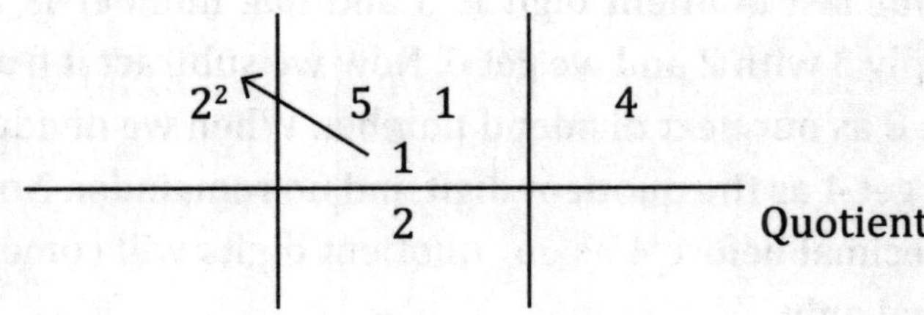

$$11 - (2 \times 2) = 7$$

So when we subtract 4 from 11, we get 7. This means our next dividend number is 7. When we divide 7 by 2, we get 3 as a quotient digit and 1 as the remainder.

2^2	5	1	$_1$4	
	2	3		Quotient

Similarly, we repeat the second step for other dividend numbers. Now 14 is the next dividend number according to the question but that is not our actual dividend number. Our actual dividend number requires that we repeat the second step here.

2^2 ↖	5	1	$_1$4	
	2	3		Quotient

$$14 - (3 \times 2) = 8$$

Our last quotient digit is 3 and flag number is 2. We multiply 3 with 2 and we get 6. Now we subtract it from 14 to get 8 as our next dividend number. When we divide 8 by 2, we get 4 as the quotient digit and no remainder. Now we put decimal before 4 as our quotient digits will come after decimal only.

2^2	5 1	4	
	2 3	4	Quotient

23.4 is the answer.

□

Solve and Practice

357 ÷ 21 = ?

689 ÷ 43 = ?

534 ÷ 12 = ?

978 ÷ 51 = ?

557 ÷ 32 = ?

Everything is a Base

Numbers not Close to Base

Till now we have learnt how to multiply numbers close to the base. As we have observed, the base is 1 with as many zeros as 10, 100, 1,000, 10,000 and so on. But not all the questions of multiplication are close to the base. So we will learn how to multiply digits not close to the base. You will be surprised to observe that we can make a base out of other numbers as well and use the same technique that we used for the multiplication of numbers close to the base. Let's take the example of 51 multiplied 52. Here we take a base either as 100/2 or 10 × 5:

$$51 \times 52 = ?$$

We take all the necessary steps assuming that the base for this particular sum is 50, though 50 is an artificial base or 50 is a base-aided base.

50 (10 × 5)

51

× 52

Now following the same pattern as the base method, we know that 51 is bigger than 50 by 1. So we write +1 on the right side of 51.

And 52 is bigger than 50 by 2. So we write +2 on the right side of 52.

50 (10 × 5)

51 | +1

× 52 | +2

First step of the base method is to multiply vertically. So when we multiply +1 with +2, we get +2 as the first part of it.

50 (10 × 5)

51 | +1 ↓

× 52 | +2

2

In the second step, we add or subtract crosswise. This means we either add 2 to 51, or add 1 to 52. In both cases, we get 53. So we write 53 on the left hand side of the answer.

```
50 (10 × 5 )
        51 | +1
            ↘
      × 52 | +2
      ---------
        53   2
      ---------
```

But 532 is not the answer. Now comes the twist, or let's say the main step to get the correct answer when the numbers are not close to the base.

As we see on the left side, 53 is written and on the right is written 2. We know that 50 is not a base; here the base is (10 × 5). So we just multiply the left side of the answer, that is 53, with 5. To get the final and the correct answer:

```
50 (10 × 5 )
        51  |  +1
             ↘
      × 52  |  +2
    -------------------
    (53 × 5 = 265) 2
    -------------------
```

So the answer is:

```
   51
 × 52
 ----
 2652
 ----
```

Let's take another example:

$$21 \times 22 = ?$$

Here in this question, both the numbers are close to 20. So we make an artificial base of 20 with support from the actual base 10. So our base will be (10 × 2).

```
20 (10 × 2)
          21
       × 22
       ----
       ----
```

21 is bigger than 20 by 1. So we write +1 on the right side of 21. And 22 is also bigger than 20 by 2. So we write +2 on the right side of 22.

```
20 (10 × 2 )
          21 | +1
               ↓
       ×  22 | +2
       ----------
               2
       ----------
```

First step of the base method is to multiply vertically. So when we multiply +1 with +2, we get +2 as the first part of the answer.

```
20 (10 × 2)
        21 | +1
          \
      × 22 | +2
      ---------
        23   2
      ---------
```

In the second step, we add or subtract crosswise, which means we either add 2 to 21, or add one to 22. In both cases, we get 23. So we write 23 as the left-hand side of the answer.

```
20 (10 × 2)
        21 | +1
          \
      × 22 | +2
      ---------
        23   2
      ---------
```

Now we know that 232 is not the answer, so we adjust the base by multiplying the left side of the answer with 2, to get the correct answer:

```
20(10 × 2 )
        21 | +1
          \
      × 22 | +2
      ---------------
      (23 × 2 = 46) 2
      ---------------
```

So the answer is:

$$\begin{array}{r} 21 \\ \times\ 22 \\ \hline 462 \\ \hline \end{array}$$

The best part of all these methods is we get all the correct answers with very little efforts. Let's check this method with another base than 10:

$$501 \times 502 = ?$$

Here we see that both the numbers are close to 500, so we take 100 × 5 as the base.

$$\begin{array}{r} 500\ (100 \times 5) \\ 501 \\ \times\ 502 \\ \hline \\ \hline \end{array}$$

When we check the difference between our base number and the numbers, we find 501 is bigger than 500 by 1. So we write +01 on the right side of 501.

And 502 is bigger than 500 by 2. So we write +02 on the right side of 502.

$$\begin{array}{rl} 500\ (100 \times 5) & \\ 501 & |\ +01 \\ \times\ 502 & |\ +02 \\ \hline & \\ \hline \end{array}$$

First step of the base method is to multiply vertically. So when we multiply +01 with +02, we get plus +02 as the first part of it.

500 (100 × 5)

```
  501 | +01
× 502 | +02
----------
        02
----------
```

Now for the second step, we add or subtract crosswise; This means we have to either add 02 to 501, or add 01 to 502. In both cases, we get 503. So we write 503 on the left-hand side of the answer.

500 (100 × 5)

```
  501 | +01
× 502 | +02
----------
  503   02
----------
```

But 50302 is not the answer. As we see on the left side it is written as 503 and on the right it is written as 02. We know that 500 is not a base. Here the base is (100 × 5). So we just multiply the left side of the answer, that is 503, with 5. To get the final and the correct answer:

500 (100 × 5)

```
       501  |  +01
  ×    502  |  +02
  ___________________
  (503 × 5 = 2515) 02
  ___________________
```

So the answer is:

```
     501
  ×  502
  ______
  251502 ------------ answer
  ______
```

Now with the help of this method, we can do multiplication of numbers which are not close to the base.

□

Solve and Practice

$$\begin{array}{r} 31 \\ \times\ 32 \\ \hline \\ \hline \end{array}$$

$$\begin{array}{r} 43 \\ \times\ 44 \\ \hline \\ \hline \end{array}$$

$$\begin{array}{r} 201 \\ \times\ 202 \\ \hline \\ \hline \end{array}$$

$$\begin{array}{r} 499 \\ \times\ 497 \\ \hline \\ \hline \end{array}$$

Playing with Dots

Multiplication of Two-Digit Number

A three-step illustration given below shows how to multiply any two-digit number with another two-digit number. These illustrations are really important to get connected with numbers.

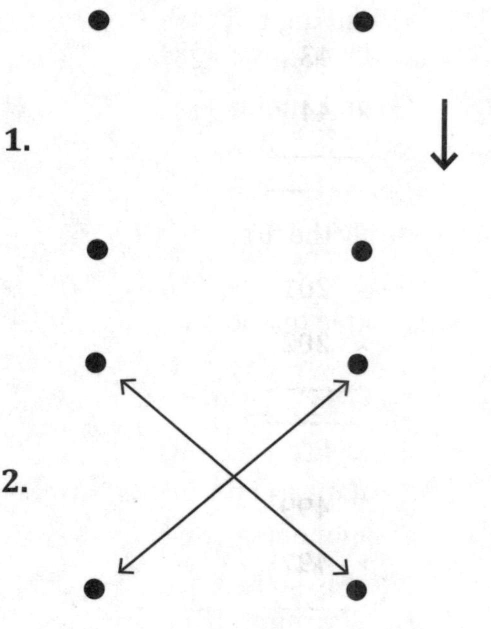

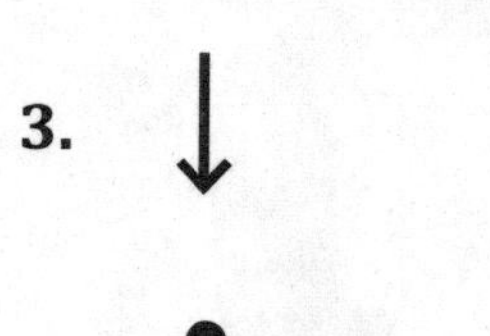

We can see that there are three steps to get the correct answer for the multiplication of any two-digit number with another two-digit number.

Step 1: First we multiply the last digits of both the numbers.

Step 2: Second, we multiply the first digit of the first number with the second digit of the second number. Then we multiply the second digit of the first number with the first digit of the second number and then add the two numbers that we got.

Step 3: We multiply the first digits of the two numbers.

Let's take an example to understand it more clearly.
Example 1:

$$21 \times 13 = ?$$

Step 1: For the last digit of the answer, we multiply the last digits of both the numbers. As we can see, 1 is the last digit of the first number 21; the second number is 13 and the last digit of 13 is 3. We multiply 1 with 3 (3 × 1 =3). This is the first step of our answer.

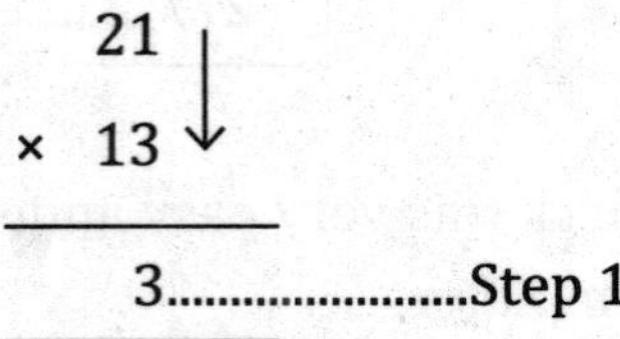

Step 2: To get the middle digit of our answer, we cross multiply 4 digits of the two numbers and then add them. So here we see that 2 is the first digit of 21 and 3 is the last digit of 13. So we multiply 2 with 3 to get 6. In the same way, 1 is the last digit of 21 and 1 is the first digit of 13. So we multiply 1 with 1. Then we add both numbers to get the second digit of our answer. Addition of 6 with 1, gives us 7, which is the second digit of the answer.

$$(2 \times 3) + (1 \times 1) = 7$$

```
   2   1
    ↖ ↗
     ╳
    ↙ ↘
×  1   3
---------
   7   3 ......................Step 2
---------
```

Step 3: We multiply the first digits of both the numbers. Here 2 is the first digit of 21 and 1 is the first digit of 13. We multiply 2 with 1 and get 2 $(2 \times 1 = 2)$. So the answer will be 273.

```
      21
     ↓
   × 13
  ______
   2 7 3.....................answer
  ______
```

Let's check this very easy and effective technique with another example.

Example 2:

67 × 42 = ???

Step 1: For the last digit of the answer, we multiply the last digits of both the numbers. As we see that 7 is the last digit of the first number 67, the second number is 42 and the last digit of 42 is 2. When we multiply 7 with 2 (7 × 2 = 14), we get 14. We keep 4 and 1 will be carried over for the previous digit. So this is the first step of our answer.

```
     67 |
   × 42 ↓
  ______
      4.....................Step 1
  ______
     1
```

Step 2: To get the middle digit of our answer, we cross multiply 4 digits of the two numbers. Then we add them. We see that 6 is the first digit of 67 and 2 is the last digit of 42. So we multiply 6 with 2 which comes to 12. In the same way, 7 is the last digit of 67 and 4 is the first digit of

42. So we multiply 7 with 4 to get 28. Then we add both the numbers to get the second digit of our answer. Addition of 12 to 28 gives us 40 and 1 is added to this number to get 41. So we take 1 as the second digit of the answer and 4 is the carry over for the previous digit.

$$6 \times 2 + 7 \times 4 = 40$$
$$40 + 1 = 41$$

```
      6     7
       ↖  ↗
        ╳
       ↙  ↘
   × 4      2
 ____________________
 (6 × 2 + 4 × 7 = 40)
     40      4
         1
     41      4   ........................Step 2
 ____________________
```

Step 3: We multiply the first digits of both the numbers. Here 6 is the first digit of 67 and 4 is the first digit of 42. We multiply 6 with 4. It gives us 24 (6 × 4 = 24). After the adjustment of carry over digit 4, we get 28 as the first number of the answer.

```
        67
      × 42
 ________________
     24  1  4
        4
     2814 .............................. Step 3
 ________________
```

```
   67
×  42
─────
 2814........................answer
─────
```

So the answer will be 2814.

□

Solve and Practice

```
   17
×  34
______

______
   63
×  81
______

______
   28
×  42
______

______
   60
×  32
______

______
   78
×  29
______

______
```

Summary

Change is always viewed with fear and negative feelings by human beings, specially when it comes to learning patterns. Only when we realise that a new way of doing the same thing that we have been doing till now can be something exciting and life-changing, then our beliefs start changing. After going through this book, we would have seen how easy and enjoyable mathematical calculations can be through Vedic Mathematics, we find to practice more similar questions and the more we practice the more confidence we gain. This also means that after reading this book, we become good friend with Mathematics.

The best way to sharpen skill in solving mathematical problems is to challenge our capabilities and go beyond the limits of our capabilities as there is always a possibility to learn more. I am sure if we all keep learning good things and adding values to our skills, we will not only help ourselves but the entire society as well.

Mathematics is a subject that is useful in our everyday life as nobody can survive in society without doing calculations. We have to do calculations in our daily life, be it for paying the servant's salary or for purchasing food items. So not only for academic purposes but even for financial transactions we have to count and sort things out. With the help of Vedic Mathematics, we can really create a difference in our calculation skills.

I am not really hopeful but sure that these techniques are helpful in your life and through your own value addition, you can help others as well. Best of luck in your daily life.

□

What is Vedic Mathematics: Q & A

Q: What is Vedic mathematics?

A: Vedic mathematics is a systematic approach to solve mathematics in very easy and small steps.

Q: Why it is called Vedic mathematics?

A: Vedic mathematics is driven by one of the four *Vedas*: the *Atharva Veda* and out of the 16 *sutras*, the entire mathematics can be solved.

Q: Who is the founder of Vedic mathematics?

A: His Holiness Jagadguru Sankaracharya Sri Bharti Krsna Tirthji Maharaja derived this exceptional knowledge of Vedic mathematics.

Q: What is the difference between Vedic mathematics and regular mathematics?

A: Vedic mathematics is the easiest approach to solving mathematics, where all other methods have certain limitations while Vedic mathematics can solve all forms of mathematics.

Q: Who should learn Vedic mathematics?

A: Vedic mathematics is useful for everyone who knows simple calculations and the best part is that learning these techniques is very easy.

Q: What practical help Vedic mathematics provides in real life?

A: Vedic mathematics is really useful in quick calculations, especially during competitive exams when a student has to answer in very little time. Vedic mathematics is very beneficial when doing any financial transactions as the calculation is required on the spot. Vedic mathematics is very useful for all students as their studies become much easier. Vedic mathematics is practically helpful in all walks of life.

Q: How to master Vedic mathematics?

A: Vedic mathematics is very easy to learn and the best way to master it is to practice – the more you practice the more adept you become.

Recently Prime Minister Narendra Modi in his 'Man Ki Baat' programme on 24 April, 2022, mentioned Vedic mathematics and how it can help students solve problems in exams, and how due to Vedic maths, calculations are made easy for every Indian.

"Via Vedic maths, you can do even the most difficult calculations in the blink of an eye."

"Vedic maths can help students solve mathematics in exams."

"Mathematics has never been a difficult subject for Indians – a big reason for this is our knowledge of Vedic mathematics."

"I wish that all parents would encourage their children to learn Vedic maths."

□

Quintana Roo, 22 February, 2014 (Mexico)

NOVEDADES | CIUDAD | QUINTANA ROO • SÁBADO 22 DE FEBRERO DE 2014 | 5

Comprueban alumnos que sí es efectiva la técnica hindú

> Estudiantes y maestros de la Técnica 28 aprenden a utilizar la capacidad de asociación para resolver problemas matemáticos

Más de 40 alumnos con bajo rendimiento escolar y ocho docentes que imparten las materias de matemáticas, geografía y ciencias básicas en la secundaria Técnica 28 "Niños Héroes de Chapultepec", ubicada en Tierra Maya, durante una semana recibieron capacitación en "Técnicas de Alto Aprendizaje", método utilizado por Manu Tripathi, investigador [illegible] lingüística y matemáticas.

Con su técnica basada en la cultura védica, el investigador mostró a sus alumnos que el proceso del aprendizaje podía ser más divertido y sencillo al ampliar el campo receptivo utilizando más de tres sentidos, de esa manera el cerebro almacena más información y logra retenerla por más tiempo, explicó.

Manu Tripathi dijo que su método busca armonizar ambos hemisferios del cerebro para una mejor comprensión [illegible] asociación. Margarita Cuevas Valencia, directora del plantel, dijo que el curso cumplió las expectativas esperadas al ver que el grupo de alumnos fueron más participativos y receptivos del conocimiento matemático.

En un examen final Manu Tripathi invitó a los alumnos a resolver ecuaciones, raíces cuadradas y multiplicaciones utilizando su "Técnica de Alto Aprendizaje", los alumnos se mostraron entusiasmados al pasar al pizarrón y hacer los ejercicios. Angel Ubaldo Loya [illegible] que Manu les mostró una forma fácil de aprender las matemáticas sin complicarse con [illegible] dían, "pude pasar al pizarrón sin temor y resolver raíces cuadradas en menos de 30 segun- [illegible] zar nuestra lógica".

Manu Tripathi invitó al grupo a repasar el conocimiento adquirido para que mejoren su rendimiento escolar en las materias que para ellos resultaban complicadas.

"Tal vez no resolví el problema completo en los métodos de enseñanza pero por lo menos con ustedes hice la diferencia y espero que utilicen esas herramientas y saquen el mayor [illegible]

AVANCES: Luego de una semana, los estudiantes con bajo rendimiento escolar pasaron al pizarrón y lograron resolver ecuaciones.

"Tal vez no resolví el problema completo en los métodos de enseñanza pero por lo menos con ustedes hice la diferencia y espero que utilican esas herramientas"

The title of this news clip in Spanish says, '*Comprueban alumnos que si es efectiva la tecnica hindu*', which translates into 'students have proved how effective Hindu techniques are.'

After completing one educational project with The Education Secretary of Quintana Roo state of Mexico by Manu Tripathi, Mexican media gave this proud moment for the country where the teachings by Manu Tripathi were recognised as a great success and credit was given to the entire country and society.

This entire book is all about how effectively and efficiently Manu Tripathi presents Vedic mathematics and other Vedic ways of learning to help students to do calculations.

बस सोचिए कि मुझे जीतना है

मनु त्रिपाठी ने बताया कि पढ़ा या सुना हुआ लंबे समय तक कैसे याद रखा जाए। जागरण

बरेली, जागरण संवाददाता : कुछ भी करने से पहले सोच लीजिए कि मुझे ऐसा कर दिखाना है। इस सोच के साथ सफलता का आधा रास्ता आप तय कर चुके होंगे। मंजिल तक पहुंचने का बाकी सफर आपकी मेहनत और कुछ बेहतर प्रबंधन से पूरा होगा।

दैनिक जागरण के तत्वावधान में आयोजित सेमिनार भूलना भूल जाएंगे में मेमोरीमैन विजय भारती और मनु त्रिपाठी ने सैकड़ो विद्यार्थियों को सफलता के मंत्र दिए। खचाखच भरे सभागार में उन्होंने कई अभिभावक, छात्रों से कई अंक और नाम पूछे, उन्हें नोट किया। इसके बाद मंच पर आए उनके साथी मनु त्रिपाठी ने ये सभी नाम और अंक उसी क्रमानुसार सुना दिए, जिस क्रम से उन्हें हाल में बैठे लोगों ने बोला था। सुनने वाले दंग थे कि कई पंक्तियों के कठिन शब्द मनु ने सिर्फ सुनकर याद कर लिए।

मंच पर आए मनु त्रिपाठी ने बताया कि हम जो करना

आज भी होगा सेमिनार

- मेमोरीमैन विजय और मनु त्रिपाठी ने दिए याददाश्त बढ़ाने के टिप्स
- होटल पंचम में दैनिक जागरण के पाठकों को निःशुल्क प्रवेश

कोशिश के साथ किया गया कार्य कठिन नहीं होता। उन्होंने वैदिक गणित के कई फार्मूले बताए जिनके जरिए गणित के सवालों को आसानी से हल किया जा सकता है। विजय भारती ने कैलकुलेशन के कई ऐसे तरीके बताए। इसके जरिए बच्चों ने कठिन गणना भी बिना कापी पेन कर दिखाई।

दैनिक जागरण में प्रकाशित सेमिनार के विज्ञापन की कटिंग लाने पर पाठकों को निःशुल्क प्रवेश मिल सकेगा। शनिवार को सुबह दस, दोपहर बारह बजे, दो

वैदिक सूत्रों से गणित के हल चुटकियों में

बरेली, जागरण संवाददाता : वैदिक गणित के सूत्रों के जरिए कठिन से कठिन सवालों के हल भी चुटकी बजाकर निकाला जा सकता है। यह दावा करते हुए मैमोरीमैन मनु त्रिपाठी ने रविवार को 'भूलना भूल जाएंगे' में तीसरे दिन भी प्रतिभागी स्टूडेंट्स को खेल-खेल में वैदिक गणित के कई सूत्र दिए।

दैनिक जागरण के तत्वावधान में आयोजित इस सेमिनार में मनु ने बताया कि वेदों में केवल नौ सूत्रों से पूरी गणित को हल करने दिखाया गया है। गणित का कोई भी ऐसा सवाल नहीं है, जिसका हल वैदिक गणित के सूत्रों के जरिए निकाला जाना संभव न हो। उन्होंने वैदिक गणित के सूत्रों के जरिए कठिन से कठिन गणित के

सवालों के हल चुटकियों में करने के टिप्स प्रतिभागियों को भी दिए। मैमोरी बेहतर बनाने के लिए आत्मविश्वास जगाने के लिए प्रेरित किया। कहा कि जब तक आप खुद को इस विश्वास से लबरेज नहीं करोगे कि

• दैनिक जागरण के तत्वावधान में आयोजित हुआ सेमिनार

हम ऐसा अवश्य कर सकते हैं। तब
वह काम संभव नहीं हो सकता। से
में मैमोरीमैन विशाल भारती ने बच्च
खेल-खेल में अच्छी मैमोरी विकसित
के टिप्स दिए।

आयोजन से जुड़े यूसी मॉस के
अग्रवाल ने बताया कि तीन दिन
सेमिनार के समापन के बाद अब 2
27 दिसंबर को पंचम होटल में वर्कश
आयोजन किया जाएगा। दोनों दिन
आठ से और अपरान्ह तीन बजे से
होगी। इसके लिए सुबह साढ़े सात
रजिस्ट्रेशन होंगे।

□□□